Energy and conservation

DATE DUE			
MAY 2 1 '91			
OCT 2 1 '91			
APR 2 2 2008			

ENERGY AND CONSERVATION

edited by ROBERT EMMET LONG

THE REFERENCE SHELF

Volume 61 Number 4

THE H. W. WILSON COMPANY

New York   1989

THE REFERENCE SHELF

The books in this series contain reprints of articles, excerpts from books, and addresses on current issues and social trends in the United States and other countries. There are six separately bound numbers in each volume, all of which are generally published in the same calendar year. One number is a collection of recent speeches; each of the others is devoted to a single subject and gives background information and discussion from various points of view, concluding with a comprehensive bibliography that contains books and pamphlets and abstracts of additional articles on the subject. Books in the series may be purchased individually or on subscription.

Library of Congress Cataloging-in-Publication Data

Main entry under title:

Energy and conservation / edited by Robert Emmet Long.
  p.   cm. — (The Reference shelf ; v. 61, no. 4)
  Includes bibliographical references.
  Summary: Discusses the present and future status of energy sources in America, focusing on natural gas, oil, solar power, and nuclear energy.
  ISBN 0-8242-0783-1
  1. Energy industries.   2. Energy conservation.   3. Energy policy.
4. Power resources.   [1. Power resources.   2. Energy conservation.]
I. Long, Robert Emmet.   II. Series.
HD9502.A2E5385   1989
333.79'16—dc20
                                                    89-16649
                                                    CIP
                                                    AC

Cover: This is the Three Mile Island nuclear power plant at Harrisburg, Pa. Reactors are in the cylindrical buildings, center. Nunber two reactor is closest to camera.
Photo: AP/Wide World Photos

Printed in the United States of America

CONTENTS

4   **The Reference Shelf**

PREFACE

Although the crisis prompted by the sharp increase in OPEC oil prices in the 1970s has passed, energy has continued to be a national concern in the 1980s. The crisis dramatized the hazards of dependency on foreign oil, but the country at present lacks a well thought-out energy policy, and new problems loom in the near future. When the high price of oil was followed by an oil glut, the U. S. oil industry suffered economic reverses and has restricted its search for new petroleum reserves. As a result, the nation is once again becoming dependent on imported oil. Moreover, there are limits to the amount of oil still obtainable in the U. S., making alternative sources of energy a prime consideration. Nuclear power, once forecast as a energy source that would not only compete with oil but eventually replace it, has proved disillusioning. The soaring costs of nuclear plant construction (and decommissioning), the question of the safety of the nuclear plants—especially following the incident at Three Mile Island, and the still unresolved problem of nuclear waste have shaken public confidence in the industry. No nuclear plants have begun construction in more than ten years, and no new plants are presently envisioned.

The emerging energy crisis, then, raises the question of where our energy will come from in the 1990s and the 21st century. Coal has supplied energy in the past, but it is a heavy pollutant of the atmosphere and a source of acid rain. Yet coal is abundant in the U. S. and with new technologies that clean the burning of coal it may play an increased role in energy production. Natural gas, once believed to exist in only limited amounts, has been discovered to be abundant below the level of four miles; and it, too, will provide a significant source of national energy supply. Among alternative sources of energy, solar energy has been making advances, and nuclear fusion may one day replace nuclear fission reactors. One of the most demonstrably successful programs in the last fifteen years has been energy conservation, which in reducing energy needs also reduces contamination of the atmosphere. American energy needs can be met in the time ahead, but the situation requires foresight and planning.

Section One of this compilation is concerned with the status of oil as a major component of the nation's energy supply and treats the aftermath of the OPEC oil crisis and present prospects for the petroleum industry. Section Two examines the nuclear energy industry and the debate over whether there may or may not be a "second nuclear era." The third section turns to discussions of natural gas, coal, and hydroelectric power, each of which offers promise of energy for the future. The fourth section deals with alternative sources, such as solar energy and nuclear fusion, as possibilities for the 21st century, as well as prospects of energy conservation.

The editor is indebted to the authors and publishers who have granted permission to reprint the materials in this volume. Special thanks are due to Joyce Cook and the Fulton Public Library staff and to the staff of Penfield Library, State University of New York at Oswego.

ROBERT EMMET LONG

July 1989

I. OIL

EDITOR'S INTRODUCTION

In the 1970s Americans were confronted by an alarming energy crisis when the OPEC countries imposed a steep increase in the price of oil. The high price of oil encouraged increased production in the U. S.; and at the same time important discoveries of oil were made elsewhere, particularly in the North Sea off Britain and in Mexico. By the 1980s, oil had become so plentiful that prices were dropping dramatically. In 1982, for example, oil sold at $40 per barrel, but by 1986 the price had declined to $15 per barrel. There was now an oil glut, a bonanza for consumers and a time of hardship for oil producers. At the end of the 1980s, the U. S. is once again importing more and more Arab oil while its own reserves are diminishing, and economists are warning that another oil crisis may be on the way. OPEC may again be raising its prices, confident that countries dependent on their oil will have to accede to their terms. Section One of this volume focuses on the vital problem of America's oil supply.

In the first article in this section, James MacKenzie, writing in *Technology Review*, discusses the certain decline of American petroleum reserves, pointing out that by 2004 the country will have consumed 90 percent of its oil. Moreover, world production of oil will begin to decline by the year 2035. The time to prepare for a transition to an alternative energy source, he argues, is now. In the following article, reprinted from *Newsweek,* Daniel Pederson comments on the disastrous effect the current oversupply of oil has had on the economy of Texas. An editorial reprinted from the *New Republic* acknowledges this hardship in the country's "oil patch," but rejects the proposal of George Bush, heading a national task force on energy, that an import fee be imposed on foreign oil. It advocates, instead, a 20-cent gas tax that would protect against a huge price hike, and would bring an additional $20 billion to the federal treasury that could be used to fill up our Strategic Petroleum Reserve.

The following selection, an editorial from *Commonweal*, notes that the National Petroleum Council's March 1987 report predicts an oil crisis in the 1990s, when the U. S. will depend on foreign sources for from 48 to 60 percent of its oil needs. The editorial faults the Reagan administration for its lack of an energy policy, advocates renewed efforts to conserve energy, and urges the immediate development of alternative sources of energy. Finally, in his article in *Fortune*, John Paul Newport similarly forecasts a dangerous dependency on foreign oil in the 1990s and calls for a new energy policy as a national priority.

LOWER OIL PRICES FUEL CONCERN[1]

The dramatic drop in world crude oil prices has provided welcome relief for consumers. But ultimately it could cause distress. That's because any related increase in consumption—which is very likely should prices stay low for several years—would hasten depletion of the world's finite amount of oil. Production could be forced to decline as early as the next decade.

This means that the need to increase fuel efficiency and develop and deploy new energy sources remains urgent. Except for its interest in nuclear fission and the Strategic Petroleum Reserve, the Reagan administration has adopted a policy of benign neglect on the energy issue. Research has been drastically cut in the belief that the market alone will solve our problems. Unfortunately, the failure to take a long-term perspective promises that the United States will be ill-prepared to deal with the inevitable transition from oil.

Despite the view of the U.S. government, the past decade has seen a growing recognition among petroleum companies and analysts that the end of the oil era is approaching. In 1982, an Exxon Corp. report on oil and gas resources stated that steady or slightly increased oil consumption would probably mean that production would reach a plateau early in the next century. The same year, Richard Nehring, a Rand Corp. analyst who has completed major reviews of global oil resources for the CIA, wrote

[1]Reprint of an article by James J. MacKenzie, senior associate at the World Resources Institute in Washington, D.C. Reprinted by permission from *Technology Review*, V. 89: 27–9. Ag./S. '86. Copyright © 1986 by *Technology Review*.

that world petroleum production would likely "decline by the early decades of the twenty-first century."

The oil price drop cannot make these forecasts rosier. After all, the decline from almost $40 per barrel in 1982 to about $15 last May was not caused by an increase in the total amount of oil that ultimately can be produced, but by reduced demand and a production increase from non-OPEC producers. Most OPEC members have also produced more than their quotas, adding to the glut.

How Much Oil on the Earth?

Geologists have made many estimates of how much crude oil can ultimately be recovered. Estimates made since the late 1970s range from 1,600 to 2,400 billion barrels. As of 1985, the location of about 1,200 billion barrels of the oil had been discovered, including about 550 billion that had been consumed. That would leave 400 to 1,200 billion barrels still to be found. Most of the undiscovered oil is probably in the Middle East, despite smaller recent finds in the North Sea, Mexico, and elsewhere. Geologists do not expect that new oil finds will change the long-term estimates.

There actually could be less recoverable crude oil than these estimates predict, according to techniques developed by M. King Hubbert, a retired U.S. Geological Survey scientist who pioneered the application of mathematical physics to geologic problems. For almost 40 years, Hubbert has analyzed data on the size of past oil discoveries and the amount of production in the continental United States to predict trends. His estimates factor in how quickly supplies are being used. To understand why his approach should apply globally, consider how it works and its success:

According to Hubbert, the data for cumulative crude oil discoveries and production can be closely fit with a simple mathematical function called the logistic curve. This s-shaped curve is derived from a bell curve that indicates annual production values at each point. The bell curve starts at zero, rises for a period of years to the maximum production level, and then falls off to zero. A logistic curve, which indicates the total amount of oil that will ever be discovered, rises and then flattens as it approaches this value.

In 1962, Hubbert plotted a logistic curve on a graph using the data for U.S. cumulative crude oil production. He predicted that production would peak in 1967, and it came in 1970—not far off the mark considering the total number of years that oil was expected to be produced. Today, with the aid of a computer, the logistic curve predicts that the total amount of crude oil that can ever be recovered in the continental United States is 190 billion barrels. It also indicates that by the year 2004 the country will have consumed 90 percent of this oil if production continues to follow the logistic curve.

Even with aggressive exploration sparked by the 10-fold price increase in the 1970s and early 1980s, domestic production has declined as Hubbert predicted it would.

That suggests that logistic curves can produce valid estimates worldwide, although far fewer discovery and production data are available for the rest of the world. Fitting a logistic curve to the total global discoveries so far yields a value of 1,450 billion barrels of ultimately recoverable crude oil. This number is obviously below the range of recent estimates. It probably would be higher if global oil production during the past 10 years had not been depressed by the high prices set by the OPEC cartel. To compensate, production data can be increased to levels that probably would have occurred under free-market conditions. Doing so suggests that there is a total of about 1,500 billion barrels worldwide—still below other recent estimates.

A variety of possible cycles for global crude oil production can also be calculated using logistic curves. If global consumption stays the same as it is now—and if there is a total of 1,500 billion barrels of recoverable crude oil—production will have to decline in about the year 2010. If all the estimates of total recoverable oil are low, and as much as 2,500 billion barrels can be recovered, production will start to decline in about 2035.

But what if lower crude oil prices result in more petroleum consumption worldwide? Then production will peak in the 1990s, assuming that there is a total of 1,500 billion barrels of recoverable crude oil. Even if there are 2,500 billion barrels of recoverable oil, the estimate moves up just slightly to a few years after the turn of the century.

Consumption Could Rise

Constraining global oil use obviously postpones the inevitable downward trend considerably and gives the world more time to develop alternative fuel sources. But global consumption is very likely to rise if oil prices stay at or below present levels through this decade. That's because oil is the cleanest, most convenient energy source. The biggest increase in oil use could occur in the developing countries and the Eastern bloc. Developing countries will probably consume more oil if prices stay low because they have little of the infrastructure needed to supply natural gas or electricity. Plus, the principal source of energy in many of these countries, wood, is being depleted rapidly. Meanwhile, the Soviet Union increased its petroleum consumption each year from 1973 to 1984, even as the West reduced its use.

Oil is attractive to the industrialized nations because it can be burned with very little pollution or other environmental side effects. While the use of more fuel-efficient heating systems and automobiles during the past decade should help matters, gasoline use will increase in the United States if people start driving more miles. That is already happening. According to the June 2, 1986, issue of *Business Week*, gas consumption had risen at an annual rate of 2.6 percent since the beginning of the year, compared with 1.8 percent growth in 1985. The United States, which consumes 25 percent of the world's oil, uses 62 percent of that amount for transportation.

Though markets will eventually adjust to declining oil supplies, prudence dictates anticipating the problem before the decline begins. Regrettably, there are few options. In the short term, an international effort must be mounted to increase energy productivity in both industrialized and developing countries. Making vehicles, houses, and industrial plants more fuel efficient would provide two to three decades of breathing room. The industrialized nations should promote this by instituting minimal efficiency standards for major energy-consuming equipment and adopting additional oil taxes. An increase in the U.S. gasoline tax would provide an incentive for consumers to purchase more energy-efficient cars. Financial and technical assistance to developing countries should focus on introducing only the most fuel-efficient technologies to avoid repeating the mistakes of the industrialized nations.

Switching to Other Fuels

In the long run, a switch to other fuels will be required. Research programs should be stepped up to accelerate the transition. Unfortunately, each of the major fuel sources that could supplant conventional oil poses problems. Coal, oil shales, and tar sands can provide plenty of energy, but their use exacerbates the risks of climate change from the buildup of carbon dioxide in the atmosphere. A massive shift to carbon-based synthetic fuels should therefore be avoided. Nuclear power issues—safety, waste disposal, and the possibility of nuclear fuel playing a role in the spread of nuclear weapons—continue to be debated. The Chernobyl accident will almost certainly slow, if not halt, the construction of more nuclear plants, at least in the West. Renewable resources, such as solar energy and wind energy, present problems because they are intermittent sources of electricity.

It is especially important to develop alternative energy sources for transportation. Hydrogen and electricity are two options to replace petroleum in vehicles, and there may be others. Expanded research in energy storage is badly needed.

The Reagan administration should focus more on these concerns if only because of the declining oil production levels in the lower 48 states. The nation is increasingly dependent on imports. Last year the United States imported 27 percent of its oil, but in 1988 it is expected to import almost half its supply. And while just a small proportion of today's imports come from the Mideast, all future additional imports will probably come from that politically unstable region.

THE GREAT TEXAS OIL BUST[2]

Texas was riding high five years ago—and when surging oil prices prompted angry talk in Washington about forcing the Lone Star State to share its oil and gas wealth with the energy-poor Northeast, true Texans

[2]Reprint of an article by Daniel Pederson, *Newsweek* staffwriter. Reprinted by permission from *Newsweek*, V. 107:16–18. M. 31, '86. Copyright © 1986 by Newsweek, Inc. All rights reserved.

responded with defiant satire. They wrote songs like "We're going to keep all the gas that we can make/And let them Yankees shiver and shake," and they decorated their cars with bumper strips like "Freeze a Yankee: Drive 75." But the jokes have faded away like the dreams of $85-a-barrel oil. Within the plush confines of Dallas's Petroleum Club, the prevailing mood today is somber. "The last joke I heard? I can't remember," says one Petroleum Club regular. "And that should tell you something."

The Texas energy boom has all but sputtered out—and the roller-coaster international oil economy is largely to blame. The combination of worldwide overproduction, continuing chaos within the OPEC cartel and reduced U.S. energy consumption has poleaxed Texas. Unemployment is on the rise, tax revenues based on oil production are declining. Bankers are nervous, wildcatters are scared and economists, charting the state's two-year plunge into recession, are scrambling to keep ahead of a free-fall decline in oil prices. In Texas, says Bernard Weinstein of Southern Methodist University, "oil is the tail that wags the whole economy." The state now loses $3 billion in gross economic output for every $1 drop in the price of oil—as well as 25,000 jobs and $100 million in state and local taxes. And the price of a barrel of West Texas intermediate crude has dropped from $25 to $13.50 over the past 10 weeks.

The signs of recession are everywhere—in the oilfields around Midland and Odessa, in the shiny emptiness of downtown Houston, even down on the farm. Texas agriculture, like farming in most other states, is in the middle of a drastic shakeout that is only marginally related to oil; but since, as Weinstein says, oil and agriculture are two of the three "legs" of the state's economy (the third is defense-related manufacturing), the result is nearly catastrophic. The worldwide oil collapse has also forced the Mexican economy to its knees—and a record tide of illegal immigration is now creating severe strain for local governments across the south Texas border region. "As long as you have less purchasing power in Mexico and more 'undocumenteds' coming across, the border area will become increasingly depressed economically," says Linda Hall, associate professor at San Antonio's Trinity University. "Nothing else is possible."

The Texas obsession with oil, in the view of many economists, is a dangerous anachronism. Oil and gas revenues now make up only 14 percent of the state's tax revenues and only 8.4 percent

of the state's revenues. Although Texas has been hard hit by the
oil industry's slump, its 8.4 percent unemployment rate is far below that of Louisiana, which is more heavily dependent on oil and
gas production. The current oil bust "was going to happen
anyway," explains Jerry Olson, associate director of the Bureau
of Business Research at the University of Texas in Austin.
"Texas's economy was based on limited resources, and they've
been fading away for a long time. It was dwindling before OPEC,
and only because of OPEC was there a major resurgence [in the
late 1970s]. Now we're back in the dwindling mode. But we had
to wean ourselves, anyway—so why not take the medicine all at
once?"

Macho spirit: The oil slump's impact on people—jobs lost,
lives disrupted, hopes deferred—is plainly worsening and there
is little doubt that the quintessential Texas brashness has been
shaken. A Texas poll published early this month showed that 51
percent of a statewide sample now consider the state an
"excellent" place to live—a 16-point decline in the past two years.
"I'm sure the downturn undermines confidence," says poll director James Dyer of Texas A&M University. "To a lot of people
here, the reason Texas wasn't suffering [during previous national
recessions] was that we had low taxes and the right macho spirit."
Dyer says many Texans simply thought "the rest of the country
wasn't doing things right," and forgot that their prosperity "was
due to a commodity-based system," particularly gas and oil.

They realize it now. Despite the fact that Texas has recorded
a net gain of nearly 400,000 jobs since 1981, the state's unemployment rate has been rising in fits and starts since 1979. It hit
7 percent in 1985—the first time in 15 years that Texas had a
higher jobless rate than the national average—and it is still going
up. The big losses, up to 200,000 jobs, have been in energy-
related industries such as oilfield-equipment manufacturing, refining and petrochemicals. Ed Spragins, president of the North
Texas Oil & Gas Association, says those job losses are undermin-
ing a way of life. "The petroleum industry in this state provided
us with a lifestyle of choice," he says. "So many people live on the
land in little old trailers on a hundred acres, working in the oilfields and enjoying the rural life-style. Now they can't sustain that
way of life."

Industry statistics tell the story. In December 1981, the number of rotary-drill rigs operating in Texas peaked at 1,448—but

by the end of 1982 that number had been cut in half. Small operators like Spragins are feeling the pinch: most of his properties are "stripper wells," which produce 10 barrels of oil or less a day. Stripper wells are shallow and inexpensive to operate, but they are also inefficient. Production costs in West Texas, for example, average $14.50 a barrel, which is higher than the current market price.

"If you go into the banks these days and say you have herpes, they say, ' Fine, have some coffee and please sit down,'" jokes oilman Dirk Stronck. "But if you come in and say you want an energy loan, they call the bouncer and throw you out."

In an obscure seventh-floor office overlooking Memorial Stadium in Austin, Bill Hamilton holds the unenviable job of budget director for Gov. Mark White. The role was as low profile as his office until State Comptroller Bob Bullock projected a $1.3 billion budget shortfall in the wake of collapsing oil revenues. White called on his 185 agency heads to trim their budgets by a hefty 13 percent—an act that, under Texas's weak governor system, can be urged but not ordered. Now Hamilton is working 12-hour days as the plans pour back in. So far, few have met the governor's goal.

Like so much of the current malaise in Texas, the state's budget problems are neither unique nor catastrophic. Louisiana and Oklahoma now face tougher economic problems by most measures. Texas grabs the attention largely because Texas is Texas—a state that, during the boom, actually held a special session to cut taxes; a state whose history and mythology have always made it a place apart, for better or worse. Political battles are shaping up over new revenue sources—a lottery, parimutuel betting or a sales-tax hike. All run against the state's religious and political tradition. And all are pushed along by the mighty force of plummeting crude prices.

"When the price of oil goes up, it generates gobs and gobs of money, and the legislature took it all and spent it," says Tony Proffitt, the state's director of tax information. "Now the legislature has to cut. That's why everybody is shaking and shivering. A job on an oil rig generates four or five jobs in the service sector and creates a healthy economy."

Houston's Shamrock Hotel was a Texas-size statement to the nation when it opened 37 years ago—as brash and bold as its founder, a legendary wildcatter named Glen McCarthy. McCarthy painted his hostelry in 63 different shades of green to honor his Irish forebears, and he installed a swimming pool big enough to water-ski in. On St. Patrick's Day, 1949, he imported 16 Pullman cars of Hollywood celebrities to open his hotel with a memorably drunken bash. The blowout was immortalized in Edna Ferber's novel "Giant"; James Dean played a very McCarthyish character named Jett Rink in the movie version of the book. But the Shamrock has a decidedly less flamboyant sequel in real life. The Hilton chain, the Shamrock's latest owners, sold the hotel to the nearby Texas Medical Center for a paltry $15 million: it will soon close and probably be demolished. On St. Patrick's Day, 1986, a still feisty Glenn McCarthy stood in the grand ballroom for the Shamrock's last hurrah. "It would be worth keeping," he said. "Even if it's just a shelter from the rain."

Houston in 1980 was a boomtown where, according to the much-repeated joke, the official bird was the construction crane—and where boomers liked to say, "You could shave an ape and he'd be a millionaire in three years." Ray Zukaitis, a 34-year-old architect who opened his own firm four years ago, remembers counting 10 major construction projects under way at the same time on a 12-block stretch along Milam Street. Houston built 43 *million* square feet of high-priced office space in 1981 alone—and today 40.3 million square feet of space is looking for tenants. Houston's real-estate bubble began to leak in 1983, along with the oil industry's. At that point, a few optimists were still saying "85 by '85," which meant that oil would be selling for $85 a barrel in 1985. But prices didn't rise, and when the bubble burst the slogan changed to "*Stay alive 'til '85.*" Now it is 1986, and the slump is still around. Ray Zukaitis is frustrated. "All the hopes for limitless growth leave you feeling betrayed," he says. "Especially if your field has anything to do with construction or oil."

Houston and its real-estate boom were heavily dependent on oil. In 1982 alone, Houston lost an estimated 100,000 jobs as drilling activity dried up. Bankruptcies rose by 33 percent in 1985—and that was after the bumper crop of 1984. While oil and gas bankruptcy filings decreased slightly in 1985, bankruptcies in the construction, contracting and development industries increased 50 percent. Retail bankruptcies increased 132 percent over 1984. The list of casualties included Sakowitz department stores, a landmark Houston chain with combined assets of more than $73 million. The chain's president, Robert T. Sakowitz, tried to turn his Houston stores into an empire that would outdo Neiman-Marcus as a purveyor of luxury goods. Like other businessmen, Sakowitz banked on rising oil prices and lost.

He wasn't the only one. Statewide thousands of Texas home buyers gambled that they could carry high mortgage rates on the strength of the boom economy: they lost, too. Residential foreclosures across the state are rising steadily, and they are especially numerous in Houston: more than 1,600 a month by one estimate, which is just about double what they were in December 1984. And with 201,000 vacant homes, Houston has a housing glut as well.

Ohioan Will Smith drove to Texas in the summer of 1976 with a brand-new M.B.A., an aging Dodge van and great expectations. He was looking for "fast action, unlimited growth . . . an environment that would grow as fast as I could produce," he says now—and for a while, he found it in Houston. Smith started out with a diesel-equipment manufacturer, then moved to a firm specializing in high-tech computer equipment for oilfield operations. The business boomed: by 1982 he was earning nearly six figures as vice president of operations, and he had started a family. But Houston's boom began to wither as oil consumption dropped—and by December of last year the bottom had dropped out. A month ago Smith took his family to Reno, Nev., to look for a new career: Smith has no regrets about leaving Texas. "The economists were really guessing wildly about when things would turn around, and the most optimistic were too long term for me," he says. "I'd be a senior citizen by the time it picked up again."

Many economists and business leaders now agree that the state must broaden its economic base. "We were caught with our pants down because we never thought of economic strategy," Weinstein says. "We have a lot of economic *growth*, but not a lot of economic *development*. We have to be innovative and competitive and entrepreneurial." That's the message being peddled by the Texas Economic Development Commission, which recently ran a full-page ad in *The Wall Street Journal* showing an astronaut wearing cowboy boots in space—a symbol of the state's new drive to marry traditional Texas business drive with high-tech and new-tech industry. Computer magnate H. Ross Perot, like Governor

White, is among the boosters of the new-tech Texas. "Oil and gas are not going away, and things are not as bad as they seem," Perot says. "Texas is not dead, Houston is not dead, and Dallas is not dead. This is a tap on our shoulder [to tell us] that we must diversify our economic base."

But there are unreconstructed Texans, like millionaire wild-catter Michel Halbouty, who are still betting on oil. "I don't give a damn what Ross Perot says or anybody else," Halbouty growls. "He's a chip man, an electronics man. People who know the oil business are not going to stand back and say there's no future in it." He adds: "We are what we are, a breed to ourselves that has always held our head up high. I know we're suffering in Texas, but we're gonna bounce back."

But when? Oil is a crapshoot with gargantuan stakes—and both the wildcatters and the forecasters have lately been very wrong in predicting the all-important price per barrel. Some economists still hope for an upturn in 1987, but those hopes are predicated on a return to higher-priced oil. The price rise may yet occur—though the OPEC oil ministers, meeting in Geneva last week, failed to agree on production limits that would make it likely. And that, in the end, is the hardest lesson of all for Texas: the state that prizes its tradition of crusty independence is captive to the volatile politics of an international cartel.

OIL DRIP[3]

The ideal price for oil would be $0 a barrel. As Henry George explained more than a century ago in *Progress and Poverty*, the cost of natural resources is nothing more than a tax on the productive elements of the economy—labor and capital. The $15 plummet in the price of oil during the past few months—from about $27 to about $12 a barrel—has been like a $90 billion annual tax cut for the American economy. The policy question is what we should do with our $90 billion.

[3]Reprint of an editorial article in *The New Republic*. Reprinted by permission from *The New Republic*, V. 194:7–8. Ap. 28, '86. Copyright © 1986 by *The New Republic*.

Vice President Bush thinks we should give part of it back. Well, who knows what—or if—George Bush really thinks. During the first week in April, Bush said regarding the proper price of oil: (a) that he was going to urge the Saudis "that we not just have a continued free-fall . . . without a parachute"; (b) that "our answer is market, market—let the market forces work"; (c) that whereas the Saudis want higher prices, the United States "ought to be pleased" by low prices; and (d) "I don't know." He continues to imply, though, that there is some danger to the nation if prices get too low. Others say it outright.

The ostensible danger of low oil prices is fourfold. First, prices could go up again if we allow ourselves to get too dependent on foreign supplies. Second, apart from price, we need the guaranteed supply of oil that only a large domestic industry can provide. Third, low prices are devastating the energy-producing regions of the country, with potential ripple effects, mostly involving the collapse of banks. And fourth, Mexico and other Third World producers are also being devastated, with potentially explosive consequences, both political and economic.

The United States is a net energy importer. Therefore, it is a mathematical certainty that the nation as a whole is better off, not worse off, with lower prices. As with any dramatic economic change, there are winners and losers, but that's capitalism. When oil prices were going up, an equally dramatic and far less desirable change, those now complaining the loudest felt very strongly that the government shouldn't interfere. As for the ripple effects, banks in trouble are forever holding themselves hostage and demanding ransom. We're skeptical. But even if a bailout were required, doing it directly would be far cheaper than showering the oil regions with money in the hope that some of it will land where it's needed.

The foreign policy consequences of cheap oil are also generally favorable. Yes, it's bad for Mexico. But it's also disastrous for the Soviet Union. Cheap oil may cost the Soviets a third of their hard-currency earnings. It's a blessing for the very poorest nations—those with no oil of their own. And it's bad for the terrorist international, whose formerly oil-rich patrons will be watching their pennies in the future.

The national security argument for maintaining a domestic oil capacity cannot be laughed off. Nor can concerns of a future OPEC rip-off if we lower our guard too much now. What can be laughed off are most of the estimates of how the domestic industry would dry up with lower prices. For one thing, prices are not likely to stay as low as they are now. If they settle at $15 to $18 a barrel, most of the most productive wells will still be profitable, and so will continued exploration. In addition, there's a countervailing factor at work. If a steel plant shuts down, you don't have a steel plant. If an oil well shuts down, you may not have an oil well, but you do have the oil—in fact, you have more of it, in more accessible locations, than if you kept pumping away full speed. When and if prices go back up, you can reopen the well. (Yes, it's not quite that easy. But it's also not as hard as oil industry apologists would have you believe.)

Of all the policies being proposed for dealing with these problems, real and imagined, the most insane is Bush's original suggestion, echoed by Secretary of Energy John Herrington, that we should talk the price of oil back up, or at least try to prevent it from falling further. To start with, this would negate all the foreign policy benefits of lower oil prices worldwide. (And surely direct aid to Mexico would make more sense than indirect aid to Mexico, the Soviet Union, and Libya.) Financially, every dollar extra we pay for oil is a 30-cent gift to foreigners and a 70-cent subsidy to domestic oil producers.

An import fee is only marginally more sensible. At least in this case the foreigners' share of the take would go to the U.S. Treasury instead. But since an import fee will raise the price for domestic oil by an equal amount, it is in effect a huge and utterly unjustified handout to domestic oil producers. A ten-dollar-a-barrel import fee, for example, would raise about $20 billion for the government and transfer another $40 billion from oil consumers (including businesses) to producers.

A gasoline tax or other broad-based energy levy would prevent this unjustified windfall. In terms of preventing renewed overdependence on foreign oil, guarding our achievements in conservation is at least as important as continued domestic production. A gas tax would have the same effect in encouraging conservation as an import fee, but for any given burden on consumers and businesses, it would bring three times as much revenue to the government. The rule of thumb is that every penny increase in the gas tax would bring in a billion dollars of revenue. A 20-cent gas tax increase, for example, would be an easy $20 billion. Gasoline prices would still be below where they were six months ago, way below what consumers pay in almost every other country, and—in real terms—even below where they were at the beginning of the energy crisis.

How to use the money? Part of it should be used to fill up the Strategic Petroleum Reserve and, if possible, to create new reserves while oil remains cheap. Huge stockpiles of quickly available oil are our best defense against either a huge price hike or a sudden shutoff by foreign producers. Buying the oil from foreigners now at $12 a barrel makes a lot more sense than paying our own producers $27 a barrel in the hope that they'll still be producing at that price when and if we need them.

The rest of the money could be used to reduce the deficit, which isn't going to go away even with all this good news. If that's intolerably sensible, it also could be used to finance a reduction in the FICA (Social Security) tax, the most regressive and economically destructive of all federal taxes, and one that has increased dramatically during the supposedly tax-cutting Reagan years.

Oh. And a few dollars might profitably be invested in a copy of *Progress and Poverty* for George Bush.

THE NEXT OIL CRISIS[4]

Remember gas lines? Hours spent waiting in a string of cars snaked around islands of gas pumps, lines that spilled out onto the street, blocking traffic. Horns blared. Drivers sighed; some cursed. One had to rev up the engine to move a foot. All this, to buy a few measly gallons of gas at recently quadrupled prices, gallons which—given the gas guzzlers most Americans drove then—wouldn't go very far.

Europeans had it even worse. Gas was rationed and use of fuel at home and at work was restricted. On brown-out days, businesses made do with neither lights, electricity-powered machines, nor heat. Stiff fines were imposed on violators. So began the winter of '73, the year OPEC became a household word.

[4]Reprint of an editorial article in *Commonweal*. Reprinted by permission from *Commonweal*, V. 114: 132-3. M. 13, '87. Copyright © 1987 by *Commonweal*.

The second oil crisis, in 1979, was milder; prices merely doubled. But virtually every year since then, dour forecasts have been published predicting the *next* oil shock. The National Petroleum Council's March 1987 report predicts the worst scenario yet, which should hit in the 1990s when the U.S. will be dependent on foreign sources for from 48 to 60 percent of its oil needs. And for good reason. Conservation and common sense are being squeezed out of current U.S. energy policy.

No sooner did OPEC agree to stabilize oil prices (around $18 a barrel) and no sooner did the price of gasoline drop to below a dollar a gallon (finally), than some Americans—under the ruse of opposing big government—developed a case of collective amnesia about gas lines and a nostalgia for V-8 engines with a speed limit worthy of them. Never mind that the Texas economy is collapsing; that "trade balance" is a euphemism; and that oil is a resource which, to the best of our knowledge, lies buried in far greater quantities in the Persian Gulf than anywhere else, our Alaskan deposits included.

Deregulation enthusiasts, led by the Reagan administration, are posing the question: What about the law? The law of supply and demand, that is. What about the consumer who *wants* to purchase a big car? What about the low-income family who *wants* a cheap (though expensive to operate) refrigerator? The logic of this position depends upon who is blamed for the last oil crisis. Check one: OPEC; the U.S. oil cartels; the U.S. government. The Reagan administration blamed the government, so now it seeks a corrective in "the unbridled market." This phrase appears to sum up our national energy policy.

Therefore, after a deadlock last term, Congress is reconsidering a bill to allow states to rescind the 55 mph speed limit. In December, Vice President Bush headed a task force which proposed: (1) to abolish the fuel economy standards, laws which require auto makers to produce energy-efficient cars or face fines (Chrysler supports the federal controls), and (2) to eliminate the Interstate Commerce Commission itself, the agency which regulates the nation's trucking and railroads.

Obviously, such a market-oriented approach promotes neither efficiency nor conservation. It does promote a course which, environmentalists predict, will accelerate the inevitable oil crash. Why take such counterproductive measures? Especially since were we to reduce our consumption of energy—as the nation had been doing with the help of legislation like the Energy Policy and Energy Conservation Act of 1975—and were able to match Western European rates of consumption, the costs of U.S. energy could be cut in half. Were we more stringent still, and able to match the Japanese consumption rates, we could extend fossil fuel reserves *for the whole world* by nineteen additional years. Such a focus could also reduce the trade deficit, reduce clean-up costs for pollutants and damage caused by acid rain, and could reduce tensions between the U.S. and the Middle East. Such a course would necessitate developing alternative sources of energy (wind, sun, water)—long out of fashion.

This laundry list of options is not innovative; it offers no immediate relief to the domestic oil industry's woes; and it does not provide a definitive solution to the world's dependence on a nonrenewable energy source. But even limited measures are better than passively waiting for the next energy crisis. The time to act is now, when at least on the energy front, the Middle East doesn't have the U.S. over a barrel.

GET READY FOR THE COMING OIL CRISIS[5]

A year of Texas-size swings in the price of crude has turned the oil patch into the trenches. Like the survivors of Verdun after the shelling stopped, the U.S. petroleum industry is spooked. True, the Organization of Petroleum Exporting Countries has apparently agreed to hold the price of oil at $18 a barrel, but the cartel has tried to agree on prices before and the troops are suspicious. Says E. H. Clark Jr., chief executive of Baker International, the giant oil field services company in California: "We're all afraid to stick our necks out of the trenches to see if the war is really over. But there's optimism. Maybe we have seen the worst."

Maybe they have. A surprising consensus has developed among petroleum experts worldwide that the benchmark Saudi light crude will sell in a range of $18 to $20 a barrel (measured

[5]Reprint of an article by John Paul Newport, *Fortune* staffwriter. Reprinted by permission from *Fortune*, V. 115:46–52, 56. Mr. 16, '87. Copyright © 1987 by *Fortune*.

in 1987 dollars) for several years to come. Prices will bob, more likely down than up, owing to the vagaries of commodities markets and the notoriously wavering discipline of the cartel members. In late February, for example, prices were slipping toward $17. But barring the outbreak of a major new Middle East war or revolution, there should be no repetition of gut-wrenching gyrations such as those of last year when oil that sold for $27 a barrel in January was fetching $10 by summer and then shooting up to $18 at year-end. Prices should be steadier than that until the early 1990s, when they will rise as world demand for crude approaches the capacity of the producing nations to supply it.

The perception that underlies this ideal view is that a mature OPEC has finally mastered the principles of basic economics. Recent years taught the cartel that insisting on top dollar for its product cost it business and hastened a market collapse. High prices really did curtail demand: The Carter Administration in 1978 foresaw petroleum use in the non-Communist world growing from 51 million barrels a day to 60 million by 1985. In fact, demand *shrank* to 46 million barrels a day in 1985.

Moreover, the high prices that OPEC maintained for oil made exploiting new fields in Alaska, Mexico, and the North Sea extremely profitable. Production in non-OPEC countries grew 28% from 1979 to 1985, topping out at 25.6 million barrels a day, a quantity nearly equal to OPEC's entire capacity. Most petroleum economists and industry executives believe that Saudi Arabia is resolved for the time being to keep oil below $20 or $22 a barrel by flooding the market if prices begin to rise. It did just that in 1986 with disastrous consequences for OPEC.

Yet another lesson learned the hard way was that if cartel members sell their oil without restraint, as Nigeria, Algeria, and others are wont to do, prices crater exactly as the economics textbooks predict. Cartel discipline may well lapse again, causing prices to plunge briefly to as low as $10 a barrel, but the recidivists are likely to reform faster in the future than they did last year. Says William M. Brown, director of energy studies at the Hudson Institute in Indianapolis and a prescient oil observer: "The experience of 1986 was a real shocker to every OPEC country, bar none. They saw that $8 oil, $7 oil, even $6 oil is possible." In January, Ecuador, an OPEC member that derives 70% of its export earnings from oil, failed to pay the interest due on its $8.1 billion of foreign debt. One of the few less developed countries that have made a point of paying their bills on time, Ecuador has been mired in economic problems resulting from the oil price collapse in the first half of 1986.

At $18, oil is high enough to sustain a profitable U.S. petroleum industry—but barely. The picture of oil that emerges in the years ahead is of an industry reduced in size, roustabout lean and rattlesnake mean, and cautious about committing huge sums for exploration and drilling. Gone will be the Rocky Mountain mobile-home boomtowns, the gaudy Texas gaucherie of Rolls-Royces converted to pickup trucks, and the fly-by-night operators drawn to oil fields by the gold rush prices of the early 1980s. The challenge for most companies will be simply to hang on for better times.

Those better times will be long in coming, however, unless the nation seriously addresses energy policy. The problem with low prices is that they encourage consumption while they discourage domestic production by rendering it uneconomic. Thus, the U.S. ultimately becomes dependent on foreign countries for its oil. The government's task should be to stimulate domestic production, encourage conservation, and spur the development of alternative sources of energy.

The real beneficiary of today's cheap oil is the U.S. economy. Although $18 a barrel is an increase of 80% from oil's summertime low, consumers will continue to get a break on their gasoline and heating bills; manufacturing costs will stay low, helping put a lid on unemployment; and the gross national product will get a healthy fillip. However, the warm narcotic of cut-rate crude is blinding the nation to the serious problem of steadily rising foreign imports.

Already the trends are apparent. From January 1986 to January 1987, U.S. crude production declined by 680,000 barrels a day, a huge 7.5% drop. Meanwhile, U.S. consumption increased by 400,000 barrels a day, growing at a faster rate than the gross national product for the first time in nine years. As a result, imports spurted by more than a million barrels a day. The American Petroleum Institute estimates that if the price stays at $18 a barrel, imports will rise from 39% of total U.S. consumption last year to more than 50% in 1991. By comparison, at the time of the first Arab embargo in 1973, imports amounted to 35% of consumption; by 1979 they accounted for 46%. Although the U.S. tries to import most of its oil from such near neighbors as Canada, Mexico, and Venezuela, in a global market, shortages and price rises affect everyone equally. And of all the oil that countries export, fully 40% comes from just five Mideast producers. Warns Chevron Chief Executive George Keller, who is the chairman of the API: "Once again our country is confronted with the possibility of a major energy crisis."

It is difficult to predict the exogenous event that would touch off such a crisis. The Iranian revolution, which provoked the 1979 shock, caught everyone flat-footed, as did the Arab oil embargo of 1973, which followed the Yom Kippur war. An Iranian victory in the Iran-Iraq conflict, however, could presage yet another oil price jolt. Iran is a price hawk. Like several other OPEC members with limited petroleum reserves, it cares little about nurturing a long-term market and a lot about forcing up prices now. A victorious Iran, with its kamikaze army and highly exportable fundamentalist religion, could easily intimidate nearby Kuwait and Saudi Arabia into cutting production to push up the price or to humble the Satan residing in the White House. Economist Alan Greenspan and others believe Iran is already pressuring the Saudis to do just that. Says he: "OPEC is now Iran–Saudi Arabia."

Tight markets increase tensions. Right now, demand in the non-Communist world for oil is about 47 million barrels a day; that's about 12 million barrels a day less than world capacity to produce it. Two-thirds of that excess capacity resides in the Middle East. But in three to five years, economic growth in the West as well as in Third World countries will sop up that surplus. Then, even the slightest Middle East contretemps that closes some spigots could seriously disrupt the economy of any nation that is highly dependent on the region for its oil. That is what worries Chevron's Keller and other industry experts.

The accord that the squabbling OPEC nations reached last December in Geneva establishes an export quota of 15.8 million barrels per day for all 13 of them. Unfortunately, that level invites cheating by such revenue-hungry members as Iran and Iraq, but the ranks appear to be holding—sort of. In February, Saudi Arabia and four American oil companies that make up the Aramco consortium signed a five-month purchase agreement at $18 a barrel. By late March or early April, demand for oil traditionally slackens, so this spring will be the first of a series of tests of the $18 mark. Notes Herbert Krupp, an oil analyst at Bankers Trust: "To maintain $18 a barrel OPEC will need to reach many, many agreements over the next few years, successfully renegotiating their quotas time after time." John Lichtblau, head of the Petroleum Industry Research Foundation, believes that if prices sustain a serious fall, OPEC members would meet quickly to readjust their quotas.

The longer OPEC can hold the line, the easier doing so will become. That is why so many economists give $18 to $20 oil a chance of prevailing. The price is high enough to provide OPEC nations with enough revenue to satisfy their immediate needs, but low enough to discourage a great deal of new production from non-OPEC sources. At $18 a barrel, oil users will not be tempted to conserve more than they do now, nor would they switch to alternative fuels as they did in the early 1980s. Thus, demand for OPEC oil could grow at about 4% per year, and members' production quotas would gradually inflate.

By 1990, OPEC will be producing at 80% of its capacity, vs. 66% today, according to a new report by the National Petroleum Council, an advisory group to the Secretary of Energy. The report notes that historically, when OPEC has reached 80% of its capacity—as it did in 1973 and 1979—the cartel "has been able to increase world oil prices and maintain them." The primary reason: Most of the excess capacity is in Saudi Arabia, Kuwait, and United Arab Emirates, three countries that rarely disagree on policy. Though the hard-earned wisdom of the early 1980s might keep the "mini-cartel" from pigging out on prices, they would yield enormous power. "Eighty percent is the danger signal," says George Mitchell, chairman of Mitchell Energy, a large production company in Woodlands, Texas. "That's when that little mini-cartel will drive you to the wall."

To a small but significant degree, the U.S. is better able to weather an oil shock than it was in the 1970s. Conservation and the exploitation of coal and other fuels have reduced oil's share of U.S. energy consumption from 47% in 1979 to 43% today. Furthermore, the 500 million barrels of oil sloshing around in hollow salt domes in Louisiana and Texas as part of the Strategic Petroleum Reserve could substitute temporarily for imported oil. But the economic costs of crude shock would still be severe; the 1979 version produced two recessions. And mere vulnerability to a jolt

would significantly hamper U.S. leverage abroad. The Reagan
Administration might have been hesitant to launch last year's pu-
nitive air strikes against Libya, an OPEC member, if world energy
markets had been tighter. Indeed, there were reports that the
Arab oil producers considered but rejected a boycott.

Declining prices have already devastated the U.S. petroleum
industry. Last summer the Hughes rig count, a measure of drilling activity, slipped below 700 for the first time since World War
II but it has climbed back to 806. In 1980, when the spot price
for the benchmark Saudi light crude was up in the ozone layer
at $39.75, there were 3,220 drilling rigs boring into U.S. rock
and seabed. Those legendary Texas wildcatters, made Midases
overnight with lucky hits, are now drilling only one or two explor-
atory wells a year, and many have gone out of the business. Typical example: oilman Clayton Williams of Midland, Texas, who
now gets more than half his diminished revenues from banking
and telecommunications.

Nearly 300,000 industry workers have been laid off since
1982, 80% of them in the oil field services industry, where four
out of ten companies have dropped out of business. "I don't think
you can afford to wipe out your staff," says Shell Oil Chief Executive John Bookout. "You may hire the bodies back, but it won't
be the same well-oiled team that works together." Today geologists step lively as bellhops in Houston hotels, and unemployed
drill hands have either drifted into other fields or resorted to
poaching game to feed their families. The stream of professionals
entering the business is drying up. Last year 192 students were
enrolled in petroleum engineering at the University of Texas,
down from 1,112 in 1982.

E. H. Clark, Baker International's C.E.O., has had to cut
more than 10,000 workers from his payrolls. "It does tear your
heart out to see what you're doing to families and lives," he says.
"Probably the hardest part is the process of closing down and selling off all the things you've built up and hired people for, and re-
alizing that a whole way of life is going to be changed."

That change is likely to be permanent. For the sad fact is that
U.S. oil production is in a sure decline. In the 80 years after 1859,
when petroleum was introduced as an alternative to whale oil—
thereby saving many a whale—the nation produced fully 64%
of all the world's oil. Because drillers always tap the largest,

cheapest, easiest-to-get-to deposits first, the ones that remain become progressively smaller and costlier to exploit. Thus, says Arco President Robert Wycoff, "even $30 a barrel was really not high enough to do anything more than barely keep production in the U.S. at a constant level. At today's prices, production will decline." Wycoff's view that this process is inevitable is widely shared. U.S. output started to decline in 1970 and continued through 1977 until Alaskan crude reversed the trend. But even that production will taper beginning in 1989, and estimates of the future slope of decline run in the range of 3% a year.

The U.S. is by far the most poked and prodded piece of landscape in the world. Geologists think nothing of plunging pipe several miles into the earth's crust looking for deposits a fraction the size of Spindletop and the great East Texas discoveries of the early decades of the century. As a result, the United States now needs 650,000 wells to pump its daily 9 million barrels of oil—not including natural gas liquids—because the average well pumps only 14 barrels a day. Five Mideast OPEC nations also produce about 9 million barrels a day, but those countries need only 3,000 wells to do the job because each pumps an average 3,100 barrels a day.

Middle East production costs are less than $1 a barrel, and the cost of finding new oil is negligible. It's a slight exaggeration to say that you could plunge a spoon into those huge deserts and the liquid would gurgle up. By contrast, U.S. production costs are roughly $7 a barrel, and finding costs, though hard to calculate with precision, are about $10, according to Arthur Andersen, the accounting firm. Says George Mitchell: "We're a high-cost country when it comes to exploration and development. The easy stuff is gone." Geology is destiny, and this is the primary reason why an industry that could make money with Saudi light selling at $1.90 a barrel in 1972, the inflation-adjusted equivalent of $4.84 a barrel today, now moans about $18.

Lower prices are forcing the industry to make some permanent cutbacks in drilling. For example, more than 90,000 of the nation's 450,000 stripper wells have been shut down since 1982. Stripper wells, owned by major oil companies as well as by tiny independent producers, pump fewer than ten barrels of oil a day and collectively supply 20% of domestic oil. But at $18 a barrel, many of these wells become uneconomic. Their income does not cover the costs of electricity to run the pumping unit, of reading the meters that monitor pressure, and of paying a royalty on the lease. Since oil and the toxic chemicals used in servicing wells can seep into and contaminate underground water supplies, most states require operators to cement up all unused stripper wells within a year. Even at higher prices, drilling through concrete to unplug the pipes, along with other start-up costs, would make re-opening most of these wells prohibitively expensive. The loss of U.S. production in the 1990s from shutting them in could run to one million barrels a day.

Financing the exploration that must be done today to yield production in two to ten years is difficult. Major oil companies are almost totally dependent on internally generated cash flow to fund their search. Last year, when prices averaged $15 a barrel, industry cash flow declined 44%, cutting exploration budgets by a third, to $26 billion. In addition, the dwindling cash flows of such major producers as Chevron, Phillips, and Unocal must be siphoned off to service high debt loads resulting from expensive mergers or shoot-yourself-in-the-foot tactics to repel raiders like T. Boone Pickens.

As a proxy for exploration, many majors are buying existing proved reserves, which is cheaper by a few dollars per barrel than finding new ones. But all they are doing is swapping assets without adding new supplies to the nation's oil cupboard. William Strevig, publisher of the *Quarterly Reserve Report* that follows such transactions, tracked a 67% run-up since 1984. "This buying and selling of resources is done by people who want to stay in business and can't justify drilling new wells," he says.

Independent oil companies, which drill about 85% of all new wells, and which must raise most of the cash they need for exploration from outside sources, have seen their traditional caches of capital dry up too. Banks, badly burned by energy loans in the past, are chary of making new ones. And the 1986 Tax Reform Act has made investing in drilling funds less attractive; investors cannot use the losses to offset ordinary income unless they have a working interest in the well, which exposes them to unlimited liability. Jeff Sandefer, at 26 a fourth-generation Texas oilman, observes that "if you take a business as risky as finding oil and gas and add price risk on top, you've given people an awfully good reason not to throw money into it."

Nonetheless, some companies continue to throw money into exploration and can find domestic oil profitably at $18 a barrel or even $15. Shell Oil, one of the most successful and best-managed U.S. oil companies, was able to replace 20% more oil and gas than it produced in 1986 at an average cost per barrel of less than $5; the company expects to do as well in 1987. Remarks C.E.O. John Bookout: "I'm not saying I'll like my return at $18 oil prices, but $18 will provide more than enough cash income to carry out the exploration program we have in mind." The company's conservative managers now premise all their exploration on selling the greasy stuff for an average of $15 a barrel. Arco, the largest domestic producer, is basing its exploration program on assumptions of $18 to $20 prices. "We can adapt ourselves to almost any price level, but it means lower profits," says Arco's Robert Wycoff. Lower profits, however, meant that Arco had to scale back its program of oil and gas exploration and engage in a $3.4-billion stock buyback program to shrink its capitalization and buoy earnings per share.

The irony is that the times have seldom been so favorable for new exploration for operators who can raise the money. Drilling costs, depressed by the industry slump, are half what they were in 1981. With fewer bidders to compete for new leases and drive up the prices, attractive properties are available on better terms. Supercomputers help geologists pinpoint likely hydrocarbon deposits with greater accuracy, cutting the number of dry holes.

When prices were high earlier in the 1980s, oil companies pioneered other techniques to detect and extract oil and gas more efficiently. When prices fell, many of these procedures became too expensive, but they could be easily used again if and when it pays to do so. MIT economist M. A. Adelman has compiled historical evidence showing that increasing knowledge about the earth's crust and advances in extraction technology, which enable companies to recover a larger portion of each deposit, have almost continuously offset the increasing difficulty of finding oil. "We will never run out of oil," Adelman says. "The real problem is the cost of providing the flow of additional reserves." Michel Halbouty, a mustachioed Houston wildcatter who seems to have been born with a divining rod in his mouth, believes that technology will be the salvation of U.S. oil. "You put these new concepts together with the right incentives and higher prices, and I believe this country could be producing ten million barrels of oil a day within ten years," he gushes. However, Halbouty's views are far more optimistic than those of most of his colleagues in the industry.

Proper exploitation of the nation's natural gas reserves could also postpone the effects of a diminishing oil supply. Natural gas takes the rap of being a scarce and declining resource. But discoveries made since the late 1970s suggest that gas resources may be many times those of oil. Although the exploration boom that followed decontrol of new finds created a gas "bubble" or surplus of some three trillion cubic feet—the nation uses 17 trillion a year—analysts expect demand will grow so long as the price of oil, to which natural gas prices are loosely tied, stays at $18 a barrel or more. Many of the nation's utilities would like to burn this clean fuel and are lobbying for the repeal of restrictive federal legislation.

The Reagan Administration has a decidedly hands-off, free-market approach to energy policy. "There is nothing we can do as a government to stop domestic petroleum production from declining," says Energy Secretary John Herrington, leaning into his words. The industry is still paying for the days of "obscene" profits and for its decisions to reinvest those profits in ways that often offended Americans. The government should not be eager to finance further forays into retailing or office equipment. However, from a public policy perspective, the national interest suffers when the U.S. is heavily dependent on foreign oil. Among the proposals that politicians, economists, and industry executives advocate for a sensible U.S. energy policy:

• **Drop the call for an oil import fee.** Enacting such a tariff on foreign crude is the most widely discussed remedy for too much dependence on imported oil. The arguments for it can be persuasive. Putting a $5 fee—a commonly cited figure—on each barrel entering the country would create a higher price for domestic producers and discourage consumption. A Department of Energy study released in April 1986 calculates that a $5 fee would reduce imports by 600,000 barrels per day and result in an extra $3.7 billion for the U.S. Treasury over five years. "The oil import fee would preserve the domestic industry so that when the price goes up they'll be there to do some extensive exploration," says Senator J. Bennett Johnston (D-Louisiana), chairman of the Senate Energy committee. "If we rely on the invisible hand, we're likely to get the shaft."

But the costs of a fee are likely to outweigh the benefits. According to the DOE study, raising the nation's cost for petroleum energy by $5 a barrel would over a five-year period boost inflation by 1% annually and reduce the growth in GNP by as much as 0.7% a year, thereby putting thousands of people out of work. More important, a fee would hurt U.S. industries that compete abroad by making them pay more than they do now for an essential raw material. Furthermore, to pass Congress, such a fee would have to have exemptions. Says the top strategic planner at a major oil company: "You'd get a Canadian exemption, a Mexican exemption, a consumer exemption. The heating oil people in Massachusetts would get preferential treatment. You'd get bureaucracy and it's counterproductive."

A gasoline tax would be far easier to administer than an import fee, since the government already collects 9 cents on every gallon that is sold. A Congressional Budget Office study concludes that imposing an additional 12 cents a gallon tax would cut oil imports by 100,000 barrels a day, would be less inflationary than an import fee, and would give the Treasury a dandy windfall of $9 billion a year. But such a tax, while salutary, would do nothing to encourage domestic oil production.

• **Open up more federal lands for oil and gas exploration.** The two most promising U.S. regions yet to be explored are the waters off the California coast, where the American Petroleum Institute estimates 3.2 billion barrels of oil could be profitably produced, and the Arctic National Wildlife Refuge, geologically next door to the 9.6 billion of proved reserves at Prudhoe Bay, which currently supply 20% of domestic U.S. oil. Unfortunately, no one can promise significant finds in either area, and environmentalists fear that oil drilling would despoil them. The annals of petroleum exploration are filled with supposed sure things like the Mukluk field in Alaska and the Baltimore Canyon off the Atlantic Coast that turned out to be billion-dollar dry holes.

Nevertheless, considering that production from offshore California and the Arctic alone might replace as much as one million barrels a day of imported oil, Congress should give industry a chance to see what it can find. The legitimate concerns of environmentalists can be protected with tough federal regulations. Says Chevron's George Keller: "Our industry's long experience in Arctic areas has demonstrated that oil exploration and development can be conducted with utmost care and protection for the natural environment." So as not to despoil the wilderness spanned by the Alaska pipeline, the builders insulated the pipe heavily to prevent the flowing crude's heat from escaping and melting the frozen tundra. The pipeline was either raised or buried as necessary to preserve caribou crossings.

Congress should also improve the terms under which it auctions leases. As it stands now, in order to win a bid on an offshore lease, companies often must pay multimillion-dollar bonuses before they know whether a leased field would yield oil. A more sensible plan might be for the government to agree to accept smaller up-front payments in exchange for a greater share of royalties if a company strikes it rich.

• **Build up the Strategic Petroleum Reserve.** The half billion or so barrels of oil now stashed in salt domes exceeds the amount of oil held back from world markets by the Arab embargo of 1973, according to John Elting Treat, executive publisher of the *Petroleum Intelligence Weekly* newsletter. The SPR's very presence discourages the use of oil as a weapon against the United States. William Hogan, professor of political economy at Harvard, estimates that once the SPR reaches its target of 750 million barrels, its oil could substitute for half of all U.S. imports for about nine months and give a jittery world situation time to sort itself out.

The Administration has proposed cutting the fill rate for the SPR in half, to 35,000 barrels a day in 1988. Representative John Dingell (D-Michigan), chairman of the House Energy and Commerce committee, has been urging the White House to keep the reserve acquisition at 75,000 barrels a day and expand the pipeline system to distribute the oil. Dingell is also a strong advocate of better contingency planning for operating the reserve and allocating petroleum in the event of a future energy crisis. "The Administration has said its policy in the event of a shut-off is to let the market take care of it," says Dingell. "I've told them at the White House that when the shut-off happens and the market *doesn't* take care of it, any fellow who comes to town with a length of rope, a pot of hot tar, and some feathers will know where to find them."

• **Repeal the windfall profits tax.** Since the last time oil companies had a few extra bucks they were apt to spend them acquiring the corporate equivalent of some very slow horses, the windfall profits tax looked like a good idea when it was passed in 1980. Oil was poised for flight into the $50 and $60 range, so the

Congress mandated a tax that at worst could take 70% of the revenues producers earned in excess of $18.50 a barrel. Because prices now are below $18.50, the tax is not generating any revenues for the Treasury and so could be repealed without any loss. "It's a bad tax," says MIT economist Adelman. "Companies don't invest in development that would otherwise be profitable."

• **Restore the depletion allowance.** This tax break permitted drillers to depreciate the value of the oil they pumped by 27.5% as they took it out of the ground. In 1975 Congress took the depletion allowance away from big integrated producers but let independent oil and gas concerns keep it, as well as other natural resource companies like copper miners and forest-product concerns. Says George Keller of Chevron: "Depreciation allowance is now a dirty word, and yet it makes eminently good sense. I don't know why your tax treatment should be different if you produce 100 barrels a day instead of producing 20,000. The economics are no different." Senator Phil Gramm (R-Texas), the balanced-budget advocate, argues that some, perhaps all, of the lost revenue caused by restoring the depletion allowance would be offset by additional taxes on the increased production it inspires. However, before the government starts handing out tax breaks to the industry, it should demand in return that any increased profits are reinvested in energy production.

• **Encourage conservation.** Lower oil prices seem to be dissipating the urgency of conservation. The President has vetoed a bill that set improved efficiency standards for household appliances, although Congress is preparing to override that veto. In 1986 the Administration decided to yield to the auto industry's pressure to roll back the manufacturers' fleet fuel-economy averages from 27.5 miles per gallon to 26 miles per gallon. Congress is debating repeal of the 55-mile-an-hour speed limit, and as a result some states are less determined to enforce it. Cutting down on gasoline consumption is especially important because some 40% of all petroleum is used by motorists, and the percentage is growing. Since there is presently no economic alternative to gas for cars, this is no time to abandon those fuel-sipping engines and a sedate speed limit.

The point of U.S. policy is to help the nation muddle through until expanded use of alternative sources of energy can make it energy sufficient—or nearly so—in the first few decades of the next century. But, "until we get to the long run, we're stuck in the short run, where all sorts of things can happen to send prices spiking up and down," says Adam Sieminski of Washington Analysis Corp., a research firm. It is also unfortunately true that the best thing for the country in the short run—low oil prices—is among the worst things for the country in the long run.

II. NUCLEAR ENERGY

EDITOR'S INTRODUCTION

Once touted as safe, reliable, and economical, nuclear power seemed the answer to a reliance on oil. But a deep sense of disillusionment with nuclear power has taken hold in the 1980s. Huge cost overruns in the construction of nuclear power plants have made this source of energy far less economical than was previously imagined. In addition the problem of nuclear waste, never satisfactorily resolved, has raised public concern. Finally, the safety of the nuclear plants themselves has been called into question by the accidents at Three Mile Island and the Soviet plant at Chernobyl. For all of these reasons the nuclear power industry is almost at a standstill. Indeed, no new nuclear plants are now being built in the U. S., and in the view of many analysts the industry is in a state of near demise.

Section Two of this compilation presents arguments for and against nuclear power. In an opening article, which appeared in *Environment*, Russ Manning reveals how elaborate safety regulations have added to the costs of building and maintaining nuclear facilities. In Manning's view, however, improvements in nuclear reactor technology may provide safe and economical plants in the future. In a following article, also from *Environment*, Dan Kirshner takes issue with Manning, arguing that a revival of the nuclear industry is unlikely. In the following article, reprinted from *The Bulletin of the Atomic Scientists*, Christopher Flavin reveals the extent of the turndown in the construction of nuclear power plants in America. As he points out, the last year a U. S. nuclear plant was ordered and not subsequently canceled was 1974. Orders for 108 reactors have, in fact, been withdrawn. Flavin surveys a number of recent reports on the future of nuclear power, which include recommendations for new, safer, and less expensive reactors; but is skeptical that a "second nuclear era" is on the way.

In an article from *Science* magazine Bernard I. Spinrad looks forward to a second nuclear era in which the present light water reactor will be replaced by an improved reactor technology. He believes that the second nuclear era could begin around 2005, with the first commercial orders of the new reactors; but he is also aware that the tide of public opinion has now turned against the expansion of nuclear energy not only in the U. S. but also in Europe and elsewhere in the world. A concluding article, reprinted from *The Nation*, discusses the much publicized case of the Shoreham Nuclear Power Station on Long Island, which was originally estimated to cost $400 million to build but came in at a cost of nearly $6 billion. Moreover, the safety of the plant is in question, "hundreds of thousands of mistakes and violations of standards" having been built into the plant. The safety issue has been particularly disturbing because the plant is located only 50 miles from New York City. When the management of the facility could produce no feasible evacuation plan, the state stepped in to prevent the plant's opening.

THE FUTURE OF NUCLEAR POWER[1]

A virtual moratorium on nuclear power now exists in the United States. Another nuclear reactor may not be ordered by a utility until after the year 2000.

During this respite, several organizations—including the Massachussetts Institute of Technology, the U.S. Congress Office of Technology Assessment, Oak Ridge Associated Universities' Institute for Energy Analysis, and the Atomic Industrial Forum—have reflected on the problems faced by nuclear power and have made proposals on how nuclear energy can be made a viable future option for electricity generation. Specifically, these studies constitute an attempt to see how nuclear power can be made more safe, more reliable, and more economical whenever utilities once again begin to order nuclear power reactors—to do it better the second time around in what Alvin Weinberg at the Institute for Energy Analysis at Oak Ridge has called the second nuclear era.

[1]Reprint of an article by Russ Manning, freelance science and environmental writer. Reprinted by permission from *Environment*, V. 27:12–17+. My. '85. Copyright © 1985 by Russ Manning.

The Current Situation

Although by 1978 utilities had already stopped ordering nuclear power plants, the first nuclear era was not destined to go quietly. On March 28, 1979, the accident at unit 2 of the Three Mile Island (TMI) nuclear plant startled the public as well as the nuclear industry and tolled the end of the first, and some say the last, attempt in this country to utilize the power of the atom to generate electricity.

Today, six years after the TMI accident and seven years after the last order for a reactor, no utility executive would dare propose to a board of directors that they begin to construct a new nuclear power plant. Too many utilities are losing their shirts on nuclear reactors.

Because they cannot manage the financial burden of the extended construction times and incredible cost overruns involved in building reactors, many utilities are stopping construction on those reactors already begun and canceling those not yet started—since 1974 over 100 reactors have been canceled. The more extravagant examples of this trend include the cancellation of the Marble Hill nuclear plant in Madison, Indiana, after $2.5 billion had already been spent; the conversion to coal of the Cincinnati Gas and Electric Company's 97 percent–complete Zimmer nuclear plant because of doubts about obtaining an operating license and because the company could not afford the additional $1.4 to $1.8 billion needed to finish it; and the default of the Washington Public Power Supply System on $2.25 billion worth of bonds after the cancellation of four of its five proposed nuclear plants.

A few reactors seem to be coming in on schedule and within a reasonable cost range. These include Florida Power and Light's St. Lucie unit 2 (constructed in six years at a relatively cheap cost of $1.4 billion) and Gulf State Utilities' River Bend reactor in Louisiana that was scheduled to begin operation last month (68 months after the start of construction and at a cost of $2.7 billion). But utilities are no longer willing to gamble on this kind of success.

Although some say that these problems are but the birth pangs of a new energy technology, many who envision a nuclear future admit that the TMI accident and the multitude of operation and management problems since have sounded the death knell of the first generation of nuclear reactors. They note that light-water reactors (LWRs), which make up the bulk of power reactors in the United States, are too complex and so are unreliable, uneconomical, and unmaintainable. They maintain that, as a result, we need a new generation of reactors (improved LWRs or completely new designs) that are safer, more reliable, and more economical.

How We Got Here

Today's LWRs are direct descendents of those first used on nuclear submarines. They were well adapted for such use because they met the essential criterion of not occupying much space. It was taken for granted that the same kind of reactor should be adapted for generation of the heat needed to run steam generators.

Unaware of latent problems, utilities showed an overwhelming initial enthusiasm for nuclear power. The genie of the atom had been released: unlimited energy was available for the asking—and many utilities asked. Small utilities, without much expertise in building, managing, or operating nuclear power plants, ordered reactors, thinking that there was not much difference between those and the coal-fired power plants they were already operating. Larger utilities, some of which had a little know-how in the nuclear field, liked the promised benefits of the initial, relatively small, unencumbered nuclear plants and quickly jumped to building extremely large plants.

Commercial reactors built in the 1960s were in the 500 to 600 megawatts of electricity (MWe) range. These included nuclear plants at Haddam Neck, Connecticut; Oyster Creek, New Jersey; and San Onofre 1, California. By the mid-to-late 1970s, the reactors coming on-line were in the 1,100 MWe range; for example, the reactor at Browns Ferry, Alabama; the Donald C. Cook reactor in Michigan; and the Zion, Illinois, reactor. This jump to large reactors occurred before adequate experience had been gained with small reactors—experience that was necessary to reveal difficulties with LWRs.

The LWR uses water to cool the core and to moderate the speed of neutrons from the fission reaction. Because the LWR is compact, it has a small thermal inertia—it changes temperature quickly. So, if anything goes wrong with the cooling of the reactor core, there is not much time to think about what to do before the core begins to heat up, later possibly melting and releasing radioactivity to the environment. (About two and one-half hours into the TMI accident, the reactor core began to melt; under other sets of conditions, melting could occur much faster.)

(New research by an ad-hoc nuclear industry group suggests significant reductions in current estimates of the amount of radioactive material likely to enter the environment in the event of a nuclear reactor accident.)

As a result of the rapid growth of nuclear power, the industry only became aware of safety problems after many reactors either had already been built and were operating or were under construction. When these safety problems came to light, the U.S. Nuclear Regulatory Commission (NRC), charged with licensing and regulating nuclear reactors, required retrofits to existing reactors and changes in design to those that were planned or already under construction. To ensure that the core never lost cooling capability, safety systems were added on, one after the other. Consequently, the LWR nuclear power plant became one of the most complicated pieces of machinery in human history.

This backward regulation created a great financial burden for the utilities. Not only did they have to bear the costs of unanticipated changes to already-operating reactors; often the reactors had to be shut down for extended periods in order to make the changes.

For reactors still under construction, changing certain designs in midstream led to extended construction times and cost overruns two to four times the original projected costs of the reactors.

In addition to high costs, the complexity of LWRs eventually led to doubts about whether anyone could understand the multitude of functions that go on, or could go on, during the operation of the reactor. One of the principal causes of the TMI accident was that the operators did not understand what was happening.

The complexity of LWRs also led to unexpected difficulties in their operation. While utilities would like their reactors to run at 70 to 80 or even 90 percent capacity, U.S. reactors in 1983 ran at an average of only 54.5 percent of annual load factor. This low-capacity factor was the result of reactor shutdowns for refueling and outages caused by problems with turbine generators, steam generators, pumps, valves, control-rod drives, and cracking pipes.

In addition to the problems caused by complexity, there were instances of mismanagement and shoddy workmanship. For example, in the Ohio Zimmer plant, the control panel could possibly catch fire in an emergency because the panel lights are too close together. The diesel generator building at the Midland plant in Michigan has settled. At San Onofre in California, the reactor was installed backwards. Reactor supports were 45° out of line at Comanche Peak in Texas. And the wrong drawings were used in assessing the seismic response for the steel in the Diablo Canyon, California, reactor containment building.

All told, these problems have led to further doubts about the reliability of plants and reports in the last few years of cost overruns that threaten to bankrupt utilities. The Seabrook nuclear plant of the Public Service Company of New Hampshire has been described as a "whopping loser"—projected at $900 million in 1972, if completed as scheduled in 1986, the two-unit reactor may cost as much as $9 billion.

Is Nuclear Power Needed?

Many nuclear power plants were ordered to meet the anticipated high rate of growth in demand for energy. But after the oil embargoes of 1973 and 1979, the American people surprised all the experts by heeding the national call to conserve energy (skyrocketing energy prices might have had something to do with it) and growth in demand for energy fell.

At present, growth in energy demand has leveled off or even fallen but electricity has continued to contribute a larger and larger part of the total energy used. Although much of the energy is lost when a lump of coal or a uranium atom is converted into electricity, electricity is attractive to many industries and other consumers because it is easily transported and is clean and efficient at the point of use.

New generating capacity will be needed in order to meet this growing demand for electricity, both to replace many oil- and gas-fired plants that are now too expensive to operate and to replace older generating units. Under the modest demand growth rate of 3 percent per year predicted by the 1983 National Energy Plan, an estimated 440,000 MWe of capacity must be installed by the year 2000, of which only 150,000 MWe is currently under construction.

Advocates of energy conservation believe that increased energy efficiency can continue to satisfy U.S. energy needs for the immediate future. But if there is no real limit to the amount of conservation that can take place, there is a practical limit to how much trouble and capital expense the public is willing and able to bear to get additional energy savings. When more electricity generation is needed, we now know that it will probably be supplied by one of four sources: coal, nuclear fission, solar alternatives, or fusion.

Because of the environmental consequences of burning fossil fuels (acid rain, strip mining, atmospheric carbon dioxide build-up), many people do not want to expand the role of coal in order to satisfy U.S. energy needs. Solar-powered electricity and fusion are still decades away from providing substantial portions of U.S. energy needs.

But will energy-efficiency measures carry the United States to a solar society decades from now? Some would argue that, if solar alternatives had received as much research and development money as has nuclear research, the United States would already be running on renewable-energy technology. Extending the argument, one could say that large research budgets continuing to go to nuclear technology will preclude advances in other technologies.

Advocates of continued work on nuclear power argue that they are not willing to risk the future of the country solely on the promise of a new, untried technology. Nuclear power should be included in the future mix of energy sources because it is tried and proved. Some plants have been built on schedule and for reasonable costs. Some plants operate at high-availability factors, and their electricity is cheaper than that from coal-fired plants.

The task before the nuclear community is to get the good record of the few to become the industry norm. In the last two years, four major studies have been completed on the future of nuclear power.

The MIT Study

The first major study on improving nuclear power came from the Department of Nuclear Engineering of the Massachussetts Institute of Technology (MIT). The report of the study, *Nuclear Power Plant Innovation for the 1990s: A Preliminary Assessment*, appeared in September 1983.

The MIT group, headed by Richard K. Lester, approached its study with the assumption that energy demand will eventually require the construction of new power plants. The first plants will be needed around 1993 with a high growth rate scenario (3.3 percent per year up to the year 2000 and 3 percent per year thereafter); around 1996 with a medium growth rate (2.3 percent per year up to 2000 and 2 percent per year thereafter); and around 2002 with a low growth rate (1.3 percent per year up to 2000 and 1 percent per year thereafter). One step to ensure that nuclear power will be able to assume its share of this new generating capacity is nuclear plant innovation.

Although many of the problems of nuclear power are beyond the control of the utilities, the MIT group believes that the worst problems have been the result of the utilities' own failures: for instance, mistakes in execution of design, mismanagement of construction projects, and poor operating practices.

But a fundamental cause of the problems in constructing and operating reactors has been the complexity of plants. "Potentially large opportunities exist," the report says, "for nuclear power plant design innovations which would reduce the demands placed on their builders and operators." Nuclear plant innovation has the potential for producing lower and more predictable plant capital costs, shorter and more predictable construction schedules, higher plant reliability, reduction of safety risk, and optimal unit sizes.

There are two technological paths away from the problems of overly complex reactors: improvements to existing LWR designs and radically different designs. The wisdom of pursuing radically different designs is still questionable. Resources for radical innovation may not be available. In addition, because of prior experience with LWRs, utilities will be inclined to favor evolutionary improvements to present designs rather than radical alternatives.

But proponents of a radical shift say incremental changes to LWR designs would not adequately address the fundamental problems of high cost and poor public acceptance; besides advocates believe that a totally new start is not necessary—much of the LWR technology is likely to be transferable. The MIT group thinks that improvements to current LWR designs will not preclude the need or desirability of investigating completely new designs that may be even better.

The group looked at various proposed new designs and found two candidates for what are called inherently safe reactors—that is, where safety resides in the reactors themselves and not in any external, redundant safety systems that must be activated in emergency situations. These are the PIUS reactor (for process-inherent ultimate safety), an LWR being developed by ASEA/ATOM, a reactor vendor in Sweden; and the high-temperature gas-cooled reactor (HTGR) developed in the United States by GA Technologies in San Diego.

The breeder reactor, the MIT group says, is not an option at present because of the high capital cost, its reliance on fuel reprocessing, and lack of the necessary technological development and operating experience to allow commercialization in this century—although small versions of liquid-metal reactors could take advantage of low-pressure operation and an effective coolant.

Of the two inherently safe designs, the MIT group favors the HTGR, especially in its modularized form. HTGRs have already been used in the Federal Republic of Germany and in the United Kingdom. In the United States, the 40-MWe Peach Bottom 1 plant in Pennsylvania operated from 1967 to 1974 and the 330-MWe Fort St. Vrain plant in Colorado has been operating since 1976.

The safety of the modular HTGR is inherent in its smaller size. If a reactor core lost coolant, the heat of radioactive decay would begin melting the LWR core in minutes, the HTGR core in hours. The core of a modular HTGR, however, would not be damaged at all, because it is so small that the heat can escape by radiation and conduction. The core will never heat to more than 1,600° C. Modular HTGRs that produce less than 100 MWe have been called walkaway reactors because, if anything goes wrong, the operator can simply walk away, since the reactor poses no danger to the public.

The inherent safety of the modular HTGR could be proved by a series of tests to demonstrate its passive safety characteristics. For example, the cooling system could be simply turned off to show that the core would not heat to dangerous levels.

In contrast, the risk would be too large to stage a loss-of-coolant accident in a conventional LWR to see if the emergency core cooling system worked. Since serious nuclear reactor accidents occur too infrequently to have any statistical significance (the TMI accident has been the only large-scale core-damage accident), the only way to judge the safety of a conventional reactor is through a process of analysis called probabilistic risk assessment (PRA).

PRA is a computer-based technique that employs "event-tree" and "fault-tree" analyses. An event-tree analysis begins with an event—a failure of a component, such as a valve or a pipe carrying cooling water—and then branches out through the reactor system from that event as from the base of a tree, looking at all the possible consequences of that failure. A fault-tree analysis begins with the end result, such as a core meltdown, and branches backward through the system to discover which failures of components would result in such an accident.

Although PRAs are difficult for someone outside the world of statistical analysis to comprehend, they are perhaps the only method—short of decades of actually running the reactors and counting the accidents that occur—by which the safety of a conventional reactor can be judged.

The PRA for a core-damage accident in a typical HTGR is 4×10^{-5} per reactor-year. That means that four core-damage accidents would probably occur in 100,000 reactor-years (50 HTGRs operating for three years would equal 150 reactor-years). This betters the NRC safety goal of 10^{-4} per reactor-year for the frequency of a core-damage accident (one accident in 10,000 reactor-years). The modular HTGR with its inherent safety would have a far lower accident frequency, perhaps zero.

The OTA Study

About five months after the MIT study was published, the U.S. Congress Office of Technology Assessment (OTA) published a report of their study, *Nuclear Power in an Age of Uncertainty*. OTA concluded that nuclear power plants currently present utilities with too many financial risks. These risks, according to the report, are the result of uncertain growth in demand for electricity, very high capital costs for nuclear plants, operating problems, increasing regulations, and growing public opposition. Of these, not much can be done about the uncertainty of growth in electricity demand, except that the development of smaller nuclear plants would allow greater flexibility in utility planning.

The paradox is that evidently these problems are not intrinsic to nuclear power, for some utilities have built and operated nuclear reactors safely, economically, and reliably. The purpose of the study, therefore, was to recommend ways in which nuclear power in general could be made a more consistently viable option.

This is not to suggest, cautioned OTA, that

nuclear power necessarily is vital to this Nation's well-being. It is, rather, to suggest that there are conditions under which nuclear power would be the preferred choice, and that these conditions might not be recognized before the industry has lost its ability to supply reactors efficiently and expeditiously. If the nuclear option is foreclosed, it should at least happen with foresight, not by accident or neglect.

High construction costs of nuclear plants can be reduced by two methods, according to the report. Nuclear plants could be built only by experienced utilities and contractors under contracts that contain incentives to control cost. In addition, the design and licensing of nuclear plants could be standardized; currently no two reactors in the United States are exactly alike and consequently the licensing requirements are different for each one.

The management and operation of nuclear reactors turned out to be much more difficult than most utilities anticipated. The OTA report suggested that technological improvements to reactors would be helpful insofar as they reduce the complexity of the plants. Nuclear service companies could be formed to help those utilities with less expertise by both building and operating the reactors.

The regulation of nuclear reactors could be changed to produce a more predictable and rational approach. Currently, proposals exist in NRC, the U.S. Department of Energy (DOE), and Congress for regulation reform; future legislation may call for standardization of plants and one-step licensing that combines the separate steps for obtaining construction and operating permits. There is still much debate over the extent to which NRC should be limited in requiring retrofits to existing reactors.

The OTA report states that public support for nuclear power is essential if nuclear energy is to remain a viable option for electricity generation. If the concerns of the public are to be addressed, there is no substitute for the long-term safe, reliable, and economic construction and operation of nuclear power plants.

The IEA Study

Soon after the OTA study appeared, the Institute for Energy Analysis (IEA), a research group within Oak Ridge Associated Universities, released a summary report of their study, *The Second Nuclear Era*, led by Alvin M. Weinberg and Irving Spiewak.

The IEA group also thinks that most of the problems of nuclear power can be traced to the complexity of the current generation of LWRs. Since the complexity of these reactors arose out of a concern for safety, and since safety is still a major factor in the public disenchantment with nuclear power, IEA sought to find a technical fix—a more simple, inherently safe reactor.

As a first step, the IEA group assessed the safety of present-day reactors, which meant looking at the changes instituted by NRC and the nuclear industry to see how much safer nuclear reactors have become since the TMI accident.

Through its *TMI Action Plan*, NRC required many changes to reactors to ensure that this type of accident would not recur. The nuclear industry responded to the accident by increasing communication between utilities and improving operation and management of nuclear reactors.

Using as a guide NRC's WASH-1400, the benchmark reactor-safety study published in 1974 that claimed that the risks from reactor accidents are very small, IEA reached the conclusion that

LWRs before TMI were more prone to accidents than the results of WASH-1400 indicated, . . . but thanks to actions taken since TMI, the frequency of accidents may have been reduced by a factor of two to three below that of WASH-1400. The net result is that LWRs today may be three to six times safer than reactors before TMI.

Although IEA found that present-day reactors meet the NRC safety goal of a core-melt probability of 10^{-4} per reactor-year, they concluded that this may not be safe enough for a second nuclear era. In a 500-reactor world, which might exist by the year 2000, the NRC safety goal would allow one core-damage accident every 20 years. In a second nuclear era, which might have many more reactors, a safety goal of 10^{-5} is probably needed.

Looking for reactors that meet these criteria, the IEA group found several designs that had core-melt probabilities lower than the NRC safety goal—for example, the HTGR with 10^{-5} per reactor-year and the Westinghouse advanced pressurized light-water reactor with 10^{-6}. Although these reactors appear to be safer because of skillful design and redundant safety systems, they would be very complex, and uncertainties about the method of PRA leave no guarantee that the reactors are actually safe. An alternative approach, according to the IEA report, is a reactor that is simpler and quite obviously safe.

Like the other study groups, IEA found the modular HTGR and the PIUS to be likely candidates for the next generation of inherently safe reactors. The IEA study favors the PIUS reactor, which at present exists only on paper and in a few demonstration experiments; there is no operating PIUS reactor.

PIUS vs. Modular HTGR

Although both the MIT group and the IEA group see advantages in pursuing either the modular HTGR or the PIUS reactor, each favors one particular design.

The MIT group likes the modular HTGR's greater thermal inertia and smaller size, which give it inherent safety; its increased safety also reduces financial risk. The reactor's small size also makes it a good candidate for standardization, for mass production in factories, and for flexibility in putting modules together to make a nuclear plant of optimal size for a given utility. In addition, HTGRs are already operating; problems still exist, but the MIT group thinks they can be resolved with about five years of research and development. Some additional advantages are the low operating power of HTGRs, which intuitively makes them safer, and the use of a gas as coolant, which leaves a cleaner cooling circuit than water and so gives maintenance workers less radiation exposure.

The MIT group has doubts about the PIUS. It thinks that all possible accident sequences have not been explored in depth. It also says that more evaluation must be done on the unique steam generation system. If a containment building is required, the PIUS, with its more expensive pressure vessel, could cost more than a conventional LWR. The complexity of the reactor could be a complicating factor in the licensing process. The complexity could also add to the expense of the reactor, and the delicate balance in the hydraulic system could cause it to shut down too frequently. In addition, although the proposed PIUS is in the 400-MWe size and several could be put together to make larger nuclear plants, it is not being proposed for very small reactors and so would not have the flexible size advantage of the modular HTGR. Also, since the PIUS is still a concept, it could take a long time to develop; the IEA group estimates it would take 10 to 12 years to have an operating plant.

Despite such problems, the IEA study came out in favor of the PIUS because of its virtually indestructible pressure vessel, which would make it inherently safe. It would be less vulnerable to sabotage and earthquake, for example.

But, in the view of the IEA group, the PIUS is more likely to be economical. The greater capital cost of the massive concrete pressure vessel could be offset by savings in the rest of the plant—since the PIUS is inherently safe, the remainder of the plant can use conventional power-plant equipment. But the IEA group recognizes that the economic success of the PIUS depends on NRC's recognizing the reactor's inherent safety, and so also recognizing that auxiliary safety systems and special containment buildings are not needed.

An additional advantage of the PIUS is that its design makes use of the light-water technology that has developed over the decades of building and operating current LWRs. The PIUS is also inherently safe even at large sizes that can take advantage of economies of scale, in contrast to the modular HTGR, which must be smaller than 100 MWe to be inherently safe.

To improve the reliability of the design, studies are under way to test the water interfaces in the PIUS and to devise methods, such as baffles at the interface regions, to stabilize the interfaces. A table model at the Tennessee Valley Authority's Water Systems Development Branch has run with stable interfaces for an hour or more with as little temperature difference as 8° C.

In fiscal year 1985 DOE has earmarked $32 million for the development of the modular HTGR. The budget contains $1.5 million for the development of advanced LWRs; the PIUS reactor could be included.

The AIF Study

A few months after the IEA and OTA studies appeared, the Atomic Industrial Forum (AIF), an international trade association, published its own assessment of the state of nuclear power, *Nuclear Power in America's Future.*

Responding to all the discussion about the promises of inherently safe reactors, AIF chided the various participants, saying they had

failed to acknowledge the extensive research, development and demonstration effort that went into alternative systems in the late '50s and early '60s . . . failed to recall the deliberative reasoning that went into the selection of the LWR . . . failed to recognize the improvements that have been incorporated into the LWR as a result of 25 years of design and operating experience . . . and finally, . . . failed to specify how they consider the LWR system to be flawed or why alternative systems could be expected to perform any better.

Although the AIF report acknowledges that research and development on LWRs and alternative reactor designs should be pursued, it suggests that current LWRs are already an inexpensive, safe, and environmentally clean method for producing electricity.

The problems with nuclear power, according to the AIF report, are the "uncertainties and financial risks associated with capital-intensive and long lead-time construction" of nuclear power plants. And the report states that the factors causing these problems are institutional, not technical.

The first priority, according to the AIF report, is safe operation of the current generation of nuclear plants. Accompanying this must be the introduction of a certain stability into the licensing and regulation of nuclear reactors, a shortening of construction times, and an increased public acceptance of nuclear power. Alternative organizational and financial arrangements might contribute to the stability of the industry: for example, fixed-price purchase contracts or incentive contracts that encourage timely construction; joint ventures by utilities to finance, build, and operate reactors of possibly standardized design; and one-step licensing that combines the construction permit and the operating license.

Where Do We Go from Here?

Perhaps the changes since the TMI accident have made nuclear reactors safe enough, and perhaps improved LWRs or totally new designs would make nuclear power even safer. But how are the American public and utility executives to be convinced of that?

"If the nuclear industry wants to survive," says Norman Rasmussen,

first and foremost . . . before breeders, before inherently safe, before anything else, [the task] is to . . . take the 80 plants we've got and invest enough time and effort to run them well and without problems, at high availability factors.

Rasmussen headed the group that produced the WASH-1400 *Reactor Safety Study* and, as a professor in the Department of Nuclear Engineering at MIT, helped produce the MIT study on nuclear power plant innovation.

Responding to question at a DOE Environment, Safety, and Health Seminar Series on probabilistic risk assessment, Rasmussen added,

I don't think making an inherently safe reactor would solve the reactor business problem at all, for several reasons. One is, it will be a long time before you convince most people that it was inherently safe. The only reason you'll have to believe it in the first place is some PRA, and most people don't understand them well enough to be comfortable with them.

IEA's Weinberg agrees that the first priority is "letter-perfect operation" of existing plants. But the second step, he says, is to move on to better reactors. "The possibility of an inherently safe reactor," Weinberg notes, "is going to haunt the utility industry until we actually build a demonstration reactor and see if it works."

But what if the country did have safely running reactors? "Do you think that's the biggest issue with nuclear power today—that there may be risks associated with them? My judgment and experience tell me no," Rasmussen says, "it's the waste disposal issue." The second reason that inherently safe reactors will not solve all of nuclear power's problems is that "an inherently safe reactor . . . will make as much waste as any other reactor."

One temporary solution to the problem of waste storage is the establishment of interim storage facilities at most existing reactor sites. IEA advocates grouping any new reactors on sites where reactors already exist, thus dedicating the sites to the long-term production of nuclear power and making it possible to store spent fuel on site for longer periods.

In 1982, Congress took a step toward resolving the problem of permanent radioactive waste storage with the passage of the Nuclear Waste Policy Act. The act established a schedule for the siting, construction, and operation of mined geologic repositories for handling high-level waste.

Although improvement in current designs, development of new inherently safe designs, plant siting, utility consortia, improved regulation, and standardization have their place in improving the nuclear industry, they all are simply ways to support the two primary necessities facing the nuclear industry in its effort to survive: the long-term, safe, uneventful, efficient, and economical operation of reactors that are already in place and the construction and operation of new reactors of whatever design; and the resolution of the nuclear waste problem.

The accomplishment of these two objectives will win over the majority of those in the public and in the utilities who have doubts about nuclear power—assuming a recognized need for new generating capacity in the first place. The question that remains is, can it be done? And since the theoretical answer to that must be yes, the real question is, *will* it be done?

DIM HOPES FOR NUCLEAR POWER[2]

Russ Manning's article, "The Future of Nuclear Power" (May 1985), is informative with regard to some of the nuclear industry's proposals for new, safer, and cheaper types of reactors with which the industry hopes to revive its economic fortunes. Despite the author's uncritical repetition of a number of arguments put forward by the industry, his article ultimately does reveal just how dim the nuclear industry's hopes are.

According to Manning, there are only "two primary necessities" facing the nuclear industry: the resolution of the nuclear waste problem, and "the long-term, safe, uneventful, efficient, and economical operation of reactors that are already in place and the construction and operation of new reactors. . . ."

Of this list of conditions, a critical one is the requirement for economical power generation. Those reactors that are already in place or under construction and have already cost more than $4

[2]Reprint of a letter by Daniel Kirshner, Environmental Defense Fund. Reprinted by permission from *Environment*, V. 27: 3–4. Jl./Ag. '85. Copyright © 1985 by the Environmental Defense Fund.

billion or $5 billion will simply not generate economic power over their lifetimes under any foreseeable circumstances. It is too late for this group of reactors to contribute anything positive to the nuclear picture.

Of course, this does not put the issue to rest. This group of nuclear plants will always have their apologists. The question remains, can the prospective economics of nuclear power improve enough to make it a viable choice for electric utilities? Apparently Manning believes this question is of interest to environmentalists for two reasons: concern with the remaining questions of the safety of nuclear power where it is viable, and concern with the environmental impacts of coal-fueled electricity generation where nuclear power is not viable.

However, neither one of these concerns is the primary reason for environmentalists to be interested in "the future of nuclear power." Concede for the moment that any new nuclear plants will run a severe safety gauntlet before construction begins. Concede also Manning's 3 percent per year growth rate in electricity use (incorrectly labeled as "modest"), which would lead to huge new plant requirements by the year 2000.

The fallacy that remains is that choices for the relevant future are limited to coal and nuclear power. Some of the country's largest utility companies are currently proving this wrong. Important alternatives like industrial cogeneration are already making major contributions in many states. Although Manning states that there is a "practical limit" to conservation without discussing what that limit is, major utilities are themselves investing in conservation, and finding ample supplies of conservation energy.

The interesting question in the future of nuclear power is how economic viability is to be achieved. The nuclear industry's instinct is to seek financial subsidies to support its goals. In proposing research and demonstration projects for new, safer, and more economic reactors, the industry looks to its biggest patron, the U.S. government. Beyond R&D, the industry seeks "Federal leadership [?] to assure adequate construction financing." The industry also has its eye on other, perhaps unwilling, contributors—utility ratepayers—in its request to "include construction work in progress in rate base" (in which case ratepayers will fund the cost of construction before a plant generates power, if it ever does). Exactly these sorts of uneconomic subsidies led to the nuclear industry's first crisis: an overcommitment to plant construction at a time when customers easily discovered more economic means to meet their end use needs through conservation. These sorts of subsidies are of concern.

To the extent that power is subsidized, mispriced, and uneconomic, there are fewer opportunities for customers to make economic choices. The economic choices will be, for a good long while yet, conservation in the form of more efficient appliances, buildings, and industrial processes, as well as alternative supplies such as cogeneration and small-scale power. The possibility that these more economic choices will be foreclosed by the nuclear industry's lobbying skills should be of the greatest concern to environmentalists.

There is a simple remedy to proposals for uneconomic subsidies. This is to enforce the discipline of the marketplace. Environmentalists would not need to oppose nuclear power if the utility industry were to bear the full economic risks of constructing, operating, and decommissioning (another topic Manning never touches) nuclear plants, and of disposing of nuclear waste. Methods to reach this end are simple: nuclear plant owners should be paid, at fair market rates, for the electricity their plants generate, and only for that power. These are the economic rules under which cogenerators and other independent power producers already live. State regulatory commissions, spurred in part by growing numbers of independent producers, are moving in the direction of applying these rules more broadly, including applying them to new nuclear plants. (The California Public Utilities Commission has instituted an investigation of the application of such pricing principles to the uncompleted Palo Verde nuclear units, of which California utilities are part owners.)

Will the nuclear industry revive? The history of attempts to create cheaper, standardized reactors is not impressive. CESSAR was Combustion Engineering's acronym for a set of standardized units, all but three of which have been canceled. The remaining three—the aforementioned Palo Verde Units—are, like so many of their nonstandardized brethren, far over budget. SNUPPS was another set, of which only two survive. Westinghouse's Offshore Power Systems had construction licenses for eight standardized units—floating reactors. None was started. The industry's attempts to develop allegedly safer High-Temperature Gas-Cooled Reactors (HTGRs) have also been marred by the cancellation of every HTGR ever proposed except the Fort St. Vrain unit, and that unit has been plagued with operational difficulties that have led it to have one of the poorest operating records of any U.S. nuclear plant. Observers of this past will not be sanguine about the future of the nuclear industry.

As long as market principles are followed in the case of nuclear energy we will all be better off. If nuclear power succeeds in passing the market test we can enjoy economic benefits. If it fails it will be because better alternatives have succeeded.

NUCLEAR POWER'S BURDENED FUTURE[3]

The Faustian bargain of nuclear energy has been lost. It is high time to leave the path pursued in the use of nuclear energy in the past, to develop new alternative and clean sources of energy supply and, during the transition period, devote all efforts to ensure maximum safety. This is the price to pay to enable life to continue on this planet.

These are not the words of an antinuclear activist but of Peter Jankowitsch, Austria's foreign minister. They were spoken not in the heat of an election campaign but before the assembled world nuclear establishment at a historic meeting of the International Atomic Energy Agency (IAEA), five months after Chernobyl. The leaders were shocked. Never before had an important member country, in this case the host of the IAEA, made such an announcement. The words are those of a politician, but they reflect a disillusionment, now widespread in Europe, with the way the nuclear enterprise has turned out.

Chernobyl has brought into sharp focus the tightening web of problems that has halted or crippled one nuclear program after another: safety, economics, waste disposal, security, and simple human blundering. The accident on April 26, 1986, may have marked the final breach of trust between nuclear leaders and their increasingly skeptical constituents. The radiologic fallout from Chernobyl was severe, but the political fallout is probably more momentous. A global march away from nuclear power may be underway.

[3]Reprint of an article by Christopher Flavin, senior researcher at the Worldwatch Institute in Washington, D.C. Reprinted by permission from *The Bulletin of the Atomic Scientists*, V. 43: 26–31. Jl./Ag. '87. Copyright © 1987 by *The Bulletin of the Atomic Scientists*.

Even before Chernobyl, nuclear power was not a healthy industry. It has long been plagued by shoddy construction practices, billion-dollar cost overruns, disinformation campaigns by government officials, thousands of tons of accumulating nuclear waste, and frequent malfunctions. Chernobyl can be compared with a heart attack that struck a patient already afflicted with cancer. And in the aftermath of Chernobyl, a growing number of political leaders appear inclined to disconnect the government-funded life-support systems that have kept the industry going so far.

The experience of nuclear planners during the past decade can be summed up in one phrase: lowered expectations. In the early 1970s, when nuclear programs were being started, many governments had lavish ambitions. In its 1974 annual report, the IAEA added up various national plans and projected that 4.45 million megawatts of nuclear power would be in place by the year 2000—more than 17 times the current nuclear capacity.

Today, the IAEA projects that the world will have just 372,000 megawatts of nuclear power in 1990 and 505,000 megawatts in the year 2000—compared with nearly 256,000 megawatts in 1986. This is almost 90 percent below the agency's 1974 projection, and a 45 percent decline from the 1980 forecast. Yet even the recent projection is more hope than forecast. Worldwatch Institute now estimates that there will be 325,000 megawatts of nuclear capacity by 1990 and 380,000 megawatts by the end of the century.

Even the supposedly successful French and Japanese nuclear programs are healthy only by the diminished standards by which such programs are now measured. Perhaps never before has an enterprise of such magnitude suffered so sudden a turn of fortune. Even optimists now recognize that the global nuclear endeavor is in deep trouble.

Despite the drastic slowdowns, however, the global nuclear enterprise is large. As of mid-1986, the world had 366 nuclear power plants in operation, with a generating capacity of 255,670 megawatts. These facilities generate about 15 percent of the world's electricity, ranging from 65 percent in France to 31 percent in West Germany, 23 percent in Japan, 16 percent in the United States, 10 percent in the Soviet Union, and none in most developing nations. Nuclear development is clearly dominated by the most economically powerful and technologically advanced nations. The United States, France, the Soviet Union, Japan, and West Germany have 72 percent of the world's generating capacity and set the international nuclear pace.

The reasons for scaling back nuclear programs are almost as diverse as the countries themselves. High costs, slowing electricity demand growth, technical problems, mismanagement, and political opposition have all had an effect. Yet these various factors actually form a complex web of interrelated problems. For example, rising costs usually represent some combination of technical problems and mismanagement, and political opposition often occurs because of safety concerns or rising costs.

The United States led the world into the nuclear age and now appears to be leading it out. The last year a U.S. nuclear plant was ordered and not subsequently canceled was 1974. Meanwhile, orders for 108 reactors have been withdrawn, representing a capacity comparable to all U.S. plants now in operation and under construction. Yet U.S. nuclear capacity continues to rise as projects started in the early 1970s finally enter operation. In 1985 and 1986, 14 plants were completed and another 19 are scheduled for completion in the next three years. After 1989, however, only three more nuclear plants are scheduled and these are all in jeopardy. U.S. nuclear capacity is likely to peak between 1990 and 1992 and will then slowly decline as aging plants are decommissioned.

The impending collapse of the U.S. nuclear construction business is driven primarily by economic forces. The rate of growth in U.S. electricity consumption has fallen from 7 percent per year in the early 1970s to 1.8 percent since 1980. While this led to a general scaling back of construction, most utility companies cut their nuclear projects first and most heavily, sometimes after spending billions on construction.

Orders and cancellations for coal plants have about balanced since 1980, while 55 nuclear projects were canceled and none ordered. The main reason for this discrepancy is cost. U.S. nuclear plants that cost less than $200 per kilowatt of generating capacity in the early 1970s carried price tags of $750 per kilowatt in 1980, $1,900 per kilowatt in 1984, and over $3,500 per kilowatt for the 14 reactors scheduled for 1987. These figures are not adjusted for inflation and so somewhat exaggerate the cost increases; but even after adjustment, they indicate a greater than sixfold increase over that period. Industry figures show that the cost of electricity from new nuclear plants (including fuel and operating expenses) now averages at least 12 cents per kilowatt-hour, while new coal plants produce at about six cents per kilowatt-hour, and gas-fired cogeneration plants at even less.

The U.S. utility industry, which in the early years was able to sweep these costs under its huge financial rug, finally has been overwhelmed by them. The industry is largely made up of private companies that must answer to investors as well as to state regulators. For many, the cost overruns were sufficient to raise prices and damage their financial condition. James Cook wrote in a February 11, 1985, article in *Forbes* magazine:

The failure of the U.S. nuclear power program ranks as the largest managerial disaster in business history, a disaster on a monumental scale. . . . It is a defeat for the U.S. consumer and for the competitiveness of U.S. industry, for the utilities that undertook the program and for the private enterprise system that made it possible. Without even recognizing the risks, the U.S. electric power industry undertook a commitment bigger than the space program or the Vietnam War.

The late 1980s may be the last gasp for the U.S. nuclear construction industry. Most of the 16 plants scheduled to begin operation during the next two years are, unfortunately, economic boondoggles that should have been canceled years ago. They are subject to extensive regulatory hearings that will determine how many of their cost overruns are the fault of the utility companies and so should be shouldered by them rather than by consumers. These last few plants will push some utilities to the edge of bankruptcy and leave their customers with high bills and bitter memories.

Cost increases have plagued nearly all the world's nuclear programs, but the increases have not been as large in most countries as in the United States. Typically, new European and Japanese nuclear plants are built for the equivalent of between $1,500 and $2,000 per kilowatt, although some in France are less expensive and some in Great Britain and West Germany cost far more. The cost of coal is higher in Europe and Japan, making it easier to justify expensive nuclear plants. Nonetheless, a late 1986 West German government study found that nuclear power's economic advantages have been substantially eroded, leaving it with almost no advantage over coal-fired plants, the direct competition.

International cost variations stem from differences in management practices and regulatory processes. Many U.S. nuclear projects are badly mismanaged, and regulation is less consistent than in other nations. Countries such as France, which have pursued standardization, have been better able to control costs. But the distinguishing characteristic of the U.S. system is its dominance by private utilities. Elsewhere, public authorities are able to quietly pass on the financial burden of nuclear power to the public at large. In the United States, nuclear power has failed a market test that it does not have to meet elsewhere.

Curiously, in other nations nuclear power is failing a political test instead—a test that Chernobyl has made more difficult. Virtually all of the world's nuclear power programs have suffered serious setbacks in recent years.

Nuclear trends in West Germany are influenced more overtly by political developments than in perhaps any other country. Since the mid-1970s that country's nuclear projects have frequently been the scene of large and sometimes violent demonstrations. West Germany's political system is relatively decentralized, with state and local governments extensively involved in the nuclear licensing process. Several projects have been stopped through a combination of public protests and legal interventions. West Germany has, in fact, started only two nuclear plants in the past decade, and the seven still to be completed include some seriously troubled projects. The West German government is considering a reprocessing plant, a fast-breeder reactor, and a high-level-waste disposal site, each of which has been subject to extensive legal actions and demonstrations.

Even before Chernobyl, there was little likelihood that West German utilities would order additional nuclear plants. Now, however, the debate is over when to shut down the country's existing plants. The ruling Christian Democratic party wants to wait until the end of the plants' useful lives; the Social Democratic party has called for a phaseout in 10 years; and the Green party, which made major gains in the 1987 parliamentary elections, wants to finish the job in two. Studies have shown that even a two-year phaseout is feasible, although electricity prices would increase slightly, as would air pollution. The future of the country's nuclear program depends on political developments, but it is almost certain that there will be no nuclear expansion during the 1990s, and there may well be a significant decline.

France's nuclear program is almost unique in having met its goals of the early 1970s. The nation already gets two-thirds of its electricity from nuclear power and in 1987 has more nuclear capacity than the next three largest European nuclear programs combined. Like much of the French system, the nuclear program is highly centralized and run by state-owned companies. Reactors are built in just six years, and neither local governments nor citizen groups have much opportunity to impede projects. Socialist President François Mitterrand came to power in 1981 promising a reevaluation of the nuclear program, but soon decided to leave it untouched. Since then, the nuclear program has been nearly immune from serious political criticism. Still, the French nuclear program faces a self-inflicted crisis: the overarching commitment to nuclear power has led the state utility, Electricité de France (EDF), to order more plants than the country needs or can afford. EDF now has a debt of $32 billion, exceeding that of most developing countries. The president of EDF admitted in 1986 that the country will have two to four "extra" reactors by 1990, and a French labor union has estimated overcapacity in 1990 at 19,000 megawatts—the output of all 16 nuclear plants ordered since 1979.

To justify its nuclear investment, EDF has shut down relatively new oil- and coal-fired power plants. It also promotes electricity consumption through bargain prices, markets power to neighboring countries, and still will have to run most nuclear plants at less than half their capacity. Even in nuclear France, however, new plant orders have slowed from six a year in the late 1970s to one a year today. The French nuclear program may turn out to be a costly relative of the Concorde supersonic jet—a technological marvel but a financial albatross.

Europe's smaller nuclear programs are also in serious disarray. Italy has faced a combination of technical problems and political opposition that for almost a decade has prevented its nuclear program from getting off the ground. A national referendum on nuclear power is now planned in Italy, and based on popular opinion so far it may well call for the end of the country's nuclear expansion program. The Spanish government canceled three nuclear plants and mothballed two others in 1983, representing a lost investment of $8 billion. Two more are being finished, but heavy reliance on nuclear power is no longer favored by the people or the government of Spain.

British nuclear expansion has slowed drastically since the late 1970s, and the country has spent five years deciding whether to abandon domestic reactors in favor of an American design. Although the official report of the lengthy Sizewell inquiry, released in early 1987, favors further nuclear expansion, as does the Conservative government, overwhelming opposition from other political parties and the public may yet stop such expansion. Meanwhile, some of the country's early reactors will probably have to be decommissioned during the 1990s. In the Netherlands, there are only two small nuclear plants in operation, and after Chernobyl the government indefinitely postponed plans to build two more.

Belgium has seven plants supplying about half the country's power, and there is not a vigorous national debate over whether to build more. Switzerland has five plants in operation, but no more are being built. Two public referenda on stopping nuclear expansion were narrowly defeated in 1980 and 1984. A third is now likely, and recent polls show that the public today is strongly against more plants.

In Finland, four reactors are operating, but no more are under construction. Since Chernobyl, opposition has become heated, and the government has suspended plans for an additional plant. In Sweden, Chernobyl increased public pressure on the Social Democratic government to fulfill the national policy, established by referendum in 1980, of phasing out all of the country's nuclear plants by the year 2010. In late 1986, a government commission determined that the phaseout could be advanced to 1997 and sharp electricity price increases avoided if the development of energy alternatives were accelerated.

Japan's nuclear power program has moved forward slowly but more steadily than most. Beyond the 33 plants that now supply 23 percent of the country's electricity, 11 are under construction. In 1984, Japan lowered the forecast for its nuclear capacity in the year 2000 by 31 percent. Japanese utilities are now ordering just two reactors per year, a rate more likely to fall than to increase. Public concern over nuclear safety was heightened by Chernobyl, but it has not translated into any effective move to stop nuclear power.

The nuclear program that is hardest to judge is the one most directly affected by the Chernobyl accident. Before Chernobyl, the Soviet Union and its East European allies had just released their latest nuclear expansion plans, which showed the Soviet Union building two-thirds of the nuclear capacity planned world-wide between 1990 and 1995 and boosting its own nuclear output by 300 percent during the next eight years. On April 25 the Soviets announced that they were canceling construction of two reactors scheduled for addition to the Chernobyl station. At the same time, they said 11 power stations were under construction elsewhere, more were planned, and that nuclear power output would reach twice the 1985 levels by 1990.

Despite these bullish projections, the Soviet nuclear program has been beset by problems that predate Chernobyl. Numerous reports have appeared of construction mishaps and delays, most serious of which are problems at the Atommash nuclear manufacturing facility in the Ukraine, designed to build equipment for all the Eastern bloc's light-water reactors. Reconstruction was required after wet ground caused the partly built plant to begin to collapse. It was completed five years later and is still not producing at the planned capacity. These and other problems have caused the Soviet Union to fall short of most of its nuclear goals by amounts that are large even by the standards of overly ambitious Soviet five-year plans.

The chairman of the Soviet State Committee on the Utilization of Atomic Energy told the IAEA September conference that the Chernobyl accident "will have no effect . . . on the further development of nuclear power generating capacity in the U.S.S.R." The restart of Chernobyl's adjacent reactors and the quick retrofits of others signal a commitment to get the nuclear program back on track, but to do this, the government has been forced to spend unprecedented resources on its nuclear program. International prestige is at stake, and Soviet officials appear to believe that they have no other way to avoid power shortages.

The Soviet Union does not change directions quickly. Reaffirmation of its nuclear goals was predictable, but there are reasons for thinking that Chernobyl will cause a reassessment. None of the seven nuclear plants scheduled to begin operation in 1986 actually did. In addition, the Soviet public is now concerned about nuclear health and safety issues, and the recent tolerance of political dissent means that these concerns may be heard. Now that the nuclear program is receiving the attention of top Kremlin officials, the extent of the problems facing it may become more apparent, and hardnosed economic judgments could replace the unrealistic goals that have driven the program so far. The Soviet Union has never been a technological leader, and the slowdown in Western nuclear programs is likely to provoke further doubts.

Before Chernobyl, a Soviet-sponsored nuclear construction program was being geared up in Eastern Europe, with no perceptible political opposition. Although East European states currently have only a few nuclear plants, a shortage of domestic energy sources and serious air pollution problems had stirred interest in nuclear power. But Chernobyl aroused considerable opposition to nuclear power in that area, and in some countries the opposition has become vocal.

Protests in several Yugoslavian republics led the government to put plans for up to eight plants on hold while a full review of the country's energy plans is carried out. Poland's latest five-year plan includes a 50 percent cut in the nuclear budget, signaling a major delay in completing that country's first four nuclear plants. East Germany and Czechoslovakia have so far indicated no change in their sizable nuclear programs. At the 1986 meeting of COMECON (the Council for Mutual Economic Assistance), six months after Chernobyl, several speeches were made in support of nuclear power, but no formal numerical commitments were issued.

Nuclear power plants in the Third World have been cut back even more drastically than in industrial countries. Most began in the 1970s, with the encouragement of foreign donors, but in recent years many of these projects have been plagued by technical problems, delays, and staggering cost overruns. Mexico's plans to build up to 20 nuclear plants have virtually collapsed. Argentina and Brazil are struggling to complete much-delayed and overbudget projects begun in the 1970s and have put further plans on hold. Brazil's sole operating plant, near Rio de Janeiro, was shut down by court order following Chernobyl and work has almost stopped at the two plants being built. The government of India has invested enormous prestige and capital in a nuclear program that is now hobbled by major technical problems and escalating costs.

South Korea and Taiwan are the only developing countries likely to get more than 10 percent of their electricity from nuclear power by the year 2000, and South Korea alone appears to be planning significant additional expansion. Once the great prospect for nuclear companies looking for export markets, China announced in March 1986 that its current plan was to build just two plants. It appears to have postponed its more ambitious earlier plans in order to avoid incurring the large debts that have accompanied nuclear programs in Brazil and the Philippines. Nuclear power may never have been economically sound for most developing nations; today it is clearly a financial disaster. For some countries, a speedy end to their nuclear programs may be a prerequisite to resolving their debt crises.

The world is now in transition from rapid growth of nuclear power in the late 1970s and early 1980s to very slow growth in the 1990s. Between 1981 and 1985, 95,800 megawatts of nuclear capacity came on line worldwide (excluding the Soviet Union and Eastern Europe), and 86,500 megawatts are scheduled for 1986–1990. But just 18,400 megawatts are scheduled for 1991–1995—a 79 percent decline in five years.

In North America and Western Europe, net additions during the 1990s will probably be lower than these figures indicate and could well be outweighed by cancellations. The wild card in international forecasting is the Soviet Union. Official Soviet plans call for 40,000 megawatts of new nuclear plants between 1986 and 1990, but even before Chernobyl it was unlikely that the target would be met. If the Soviet nuclear program instead proceeds at its 1985 pace of five plants a year, global nuclear capacity could peak before the end of the century as the retirement of aging plants exceeds new generating capacity. Even that pace may not be sustained and the more the Soviet program is set back, the sooner the global peak will come.

Nevertheless, the weight of official policy in many nations is still behind nuclear power. The head of the Soviet delegation to the 1986 IAEA meeting said that "the exploitation of the atom's energy has become a realistic requirement, and is preconditioned by interests of human civilization progress." West German Chancellor Helmut Kohl told a national television audience: "Abandoning nuclear power could spell the end of the Federal Republic as an industrialized nation." Energy Secretary Peter Walker of the United Kingdom said: "If we care about the standard of living of generations yet to come, we must meet the challenge of the nuclear age and not retreat into the irresponsible course of leaving our children and grandchildren a world in deep and probably irreversible decline."

The leaders of three of the world's most economically powerful nations could hardly state their faith in nuclear power more forcefully. Yet a growing number of national leaders take the opposite position. Chernobyl sparked the formation of a "non-nuclear club" composed of countries with the capability and past inclination to rely on nuclear power, but which have decided to forgo this source of energy. Several countries, including Australia, Denmark, and New Zealand, made deliberate decisions in the early 1980s not to develop nuclear power, and Sweden made plans to phase it out. Chernobyl led Austria's government to decide to dismantle the country's only reactor at Zwentendorf. Corazon Aquino, president of the Philippines, announced soon after Chernobyl that the newly completed Bataan nuclear plant would be dismantled. Greece's socialist government abandoned plans for that country's first reactor.

The non-nuclear club may soon gain some new members. Although the ruling government in West Germany favors nuclear power, the opposition Social Democratic party is pushing to close the country's plants. The future of the nuclear program there will depend on political developments, but according to Britain's influential *Financial Times* (January 14, 1987), "signs are multiplying that a long slow German march away from atomic energy may already be under way." In Italy, a national referendum is planned for this year to decide the fate of the country's nuclear program. Switzerland and Britain appear to be moving slowly toward similar decision points.

The further question—what will replace nuclear power—remains largely unaddressed in most countries, although some discussion has begun. Many West European countries are re-evaluating the potential for increasing energy efficiency and developing alternative energy sources. The West German Greens and Social Democrats think it possible to phase out nuclear power in 10 years or less, in part because West German generating capacity now outruns demand by some 20 percent. But in the long run new sources will be necessary and both West Germany and Greece have reinvigorated their renewable energy programs since Chernobyl. The Swedish government commission recommending a 10-year phaseout also called for increased investment in energy alternatives.

As a practical matter, it is questionable whether countries can or should base their power systems on technologies that much of the public finds objectionable. Modern societies are confronted with many controversial issues, but the nuclear endeavor is almost unmatched in the steadily mounting resistance it faces. In the United States, opposition to building nuclear power plants rose from 19 percent in 1975 to 78 percent (following Chernobyl) in 1986.

A similar shift appears to have occurred in Europe. Although Three Mile Island and Chernobyl stirred public opinion, there is more at work here. Simply put, many people believe that nuclear power programs have gone badly awry, they do not trust officials to fix the problems, and they would rather rely on some other energy source. Public opinion now differs profoundly from official policy in many countries, and over time these differences may begin to tear at the national political fabric. The political cost of government pro-nuclear campaigns could well exceed potential benefits.

The extent of opposition to nuclear power may also reflect a realization of the ultimate implications of pursuing the nuclear path. If the plans laid in the early 1970s were realized, by the year 2010 several thousand fission reactors would be operating, many at the edge of large cities. Hundreds of breeder reactors would be fueled by plutonium, transported along the world's highways and railroads from the reprocessing plants where it had been refined. Nuclear power plants, fuel manufacturing facilities, and waste sites would be as common in 2010 as steel plants are today. Without a major improvement in industry performance, serious nuclear accidents would be regular occurrences in the twenty-first century.

The world's nuclear leaders have lost their early vision but have not found a new one. In the aftermath of Chernobyl, there is an almost overwhelming feeling among the general public that the current generation of nuclear technology is simply not working. This deep-seated disenchantment can only presage a continued decline for nuclear power. For energy, as for all important issues in democratic societies, the final decision must be made, as Albert Einstein put it, "in the village square." Energy is, at base, a political issue. In this context, nuclear power's prospects have never been dimmer.

Chernobyl's true cost remains unknown, but its challenge is clear. Nations must set about developing a new energy vision, to carry them beyond the divisive, often debilitating, struggle that has come to seem inherent in the very word "nuclear."

U. S. NUCLEAR POWER
IN THE NEXT TWENTY YEARS[4]

From 1965 to 1975, the use of nuclear power in the United States grew dramatically. However, after 1975 this growth slowed down, and it has now almost stopped. New nuclear power plants are too expensive, and the public does not have confidence in them. Therefore, the product must be improved if it is to be salable. Work toward this end is in progress. The service life of existing nuclear plants must also be extended. These two developments permit some cautious estimates of the status of nuclear power 20 years from now. From about 1990 to 2005 there will probably be a pause in the installation of nuclear plants. During that period the nuclear community will be developing and testing new reactor designs for economic and demonstrably safe nuclear power.

The travail of the nuclear industry began in about 1975, when it became obvious that the shock of the oil price set by the Organization of Petroleum Exporting Countries had brought about an industrial recession and a sudden leveling of demand for electricity. Utilities, faced with unnecessary capital commitments, curtailed the purchase of new generating plants. Nuclear plants suffered the most, because most of the plants on order were nuclear and because nuclear power is capital-intensive.

Nevertheless, nuclear power remained economically attractive until 1979. Its generating cost was less than that of coal power, and, in refutation of public opposition, it had an excellent safety record. Therefore, the nuclear industry expected orders to resume as soon as electrical demand began to climb again.

[4]Reprint of an article by Bernard I. Spinrad, professor and chair, Department of Nuclear Engineering, Iowa State University. Reprinted by permission from *Science*, V. 239: 707–8. F. 12, '88. Copyright © 1988 by the American Association for the Advancement of Science.

Then the accident at Three Mile Island (TMI) unit 2 occurred. In the aftermath of TMI, major changes were made in nuclear regulations. The number of nuclear plant personnel and the training required of them were greatly increased; extensive backfitting of hardware was also required.

These measures increased the cost of nuclear power, but initially the increase was expected to be modest. However, in the 1980s, plant capital costs escalated. The cost of new nuclear plants reached three to five times what had been originally expected. The price charged for electricity normally includes a "reasonable" return on investment, but at standard rates of return these high costs caused sudden jolts in electricity rates. Some utility commissions began to disallow costs judged to be excessive in setting rates, forcing these costs to be absorbed by stockholders. Only a few, very well-managed nuclear utilities were able to avoid these embarrassments, and nuclear power lost its support in the utility sector.

Many theories as to the cause of this cost escalation have been propounded, most of which blame factors outside the nuclear industry. My theory is that the cause is within the industry. The plant designs now being offered are large, costly, and complicated. Regulatory requirements have been patched into plant designs rather than incorporated into the basic design. The nuclear industry did not see any long-term profit in going beyond incremental design changes for the existing types of nuclear plants. Under these circumstances, the enthusiasm of the design staff waned. Errors began creeping into design and construction, requiring extensive reworking. These problems were combated by hiring larger staffs to do more checking and rechecking, inspection, and paperwork. Nuclear power plant design and construction have become highly bureaucratized—and more expensive.

Nuclear power has also lost its base of popular support. Public confidence plummeted after TMI, and recovery of confidence was frustrated by the Chernobyl accident. A nuclear project will be strongly fought at all governmental levels. Public interventions and lawsuits will delay the project. The public image of the utility will be hurt, which will jeopardize the utility's case in other disputes. Under these circumstances, the utility can only justify the project to its stockholders and the public if it can guarantee that cheap electricity will be produced; but the recent cost history of nuclear plants does not make such a guarantee feasible. Nuclear power will not be a significant factor in new electricity generation unless or until its economy can be ensured.

A "second nuclear era" seems to be needed if nuclear power is to make a growing contribution to generation of electricity. The term was coined by Weinberg to denote a period when nuclear power is again publicly accepted and cheap. The environmental benefits of nuclear power are compelling, and if they are associated with both economy and safety they will be hard to oppose. Public response to safety issues becomes muted if the industry being questioned maintains a good safety record for a decade or so. These, then, are the conditions for the second nuclear era. The questions are what types of systems will characterize it, and when will it begin.

The prevailing view in nuclear circles is that new reactor designs will be needed, but no new reactor types. Types in use today—the light water reactors, the heavy water reactor, the sodium-cooled fast reactor, and the helium-cooled graphite reactor—were selected for valid reasons, and they are the points of departure. They all have attractive intrinsic safety features, but concerns about safety will not be appeased unless safety is clearly apparent. Thus, the design task is to provide transparent safety and economic construction and operation.

One key to economy may be to decrease the size of individual reactor units. This amounts to abandoning economy of scale in favor of economy of standardization and replication. Smaller units would also (i) limit the cost exposure of each incremental unit, (ii) provide a better match to growth of the utility load, and (iii) take greater advantage of the economies and cost management of factory construction, as opposed to field construction. A precedent for smaller units is found in the unit sizes of fossil-fueled plants, which are typically built in the 250- to 400-MW range, and hardly ever exceed 800 MW.

Another change in philosophy concerns automation. Nuclear plants have been designed for a high degree of human involvement in operation. This is no longer the practice in modern manufacturing, and it is not considered conservative in the sense of safety. Now, the nuclear industry is also starting to consider automation, particularly for situations that do not permit the exercise of operator judgment. Moreover, utilities are installing a variety of computerized, on-line expert systems to guide both maintenance and operations. These automation changes should improve both economy and safety: economy by reversing the trend toward increased staffing, safety by removing a demonstrated weak link—as illustrated by both TMI and Chernobyl—from many of the safety chains.

Another technical opportunity comes from robotics. Human surveillance and servicing of many plant components are costly in terms of radiation exposure and money. The exigencies of TMI led to the use of robots in inspection and cleanup after the accident. These experiences were successful, and plant design for optimum use of robots and teleoperators is now in favor. These techniques will be useful, but the second nuclear era must be based on better reactors. Design programs are under way for several types.

The workhorse reactor type today is the light water reactor (LWR). Advanced LWRs are now under design. They are likely to be smaller (about 600 MW, compared to today's reactors of 1200 MW), more durable (60-year design life rather than 30 years), more conservative with larger safety margins, and, above all, simpler to make proof of safety less equivocal.

Advanced fast reactor design and development is also under way. A consensus has developed in favor of the pool type with all primary components in a single, low-pressure sodium pool, a low power of less than 500 MW per unit, modest breeding, and moderate specific power. Metal, rather than oxide, fuel is now preferred, on the basis of experiments at Argonne National Laboratory that showed metal fuel to have impressive safety characteristics. These characteristics permitted the Experimental Breeder Reactor (EBR-II) to survive severe operational accident simulations without damage to fuel or major stress on the system. The pool concept and small size permit emergency cooling by natural circulation of building atmosphere (possibly nitrogen) around the pool tank. Illustrative designs by General Electric and Rockwell International have been published.

Proponents and developers of helium-cooled high-temperature reactors (HTGR) in both the United States (General Atomic) and West Germany were the first to downsize their concept to small modular units, making conventional pressure vessels and emergency cooling by circulation of air in the building feasible. Finally, the Canadian heavy water reactor line, known as CANDU, could possibly be adapted to U.S. practice; these reactors have compiled excellent operational records.

These advanced and alternative reactor types will need to have their safety, operability, and economy tested by experimental construction and operation. However, with so many possible reactor types, I expect at least one of them to pass this test and to be commercialized. The small unit size of the final product is important in this process, since it would require much less time, money, and extrapolation to go from an experimental plant to a commercial line.

Success in at least one of these programs is the real key to a second nuclear era. I anticipate such a success, but it will take time. A scenario for the process might be as follows: Experimental construction begins in 1992; experience justifies a prototype plant in 1999; and commercial orders begin around 2005. This scenario puts the second nuclear era a full human generation into the future. Most large-scale industrial innovations are realized over that length of time. It took that long for the first nuclear era, which is tailing off now, to bloom, and our present circumstance is indeed one of beginning again.

In the interim, many existing nuclear plants will reach the end of their license periods. They will still be in generally excellent condition, which is a requirement as long as they are licensed. Some components will be at the end of their useful lives and will need refurbishing or replacement, but most of the equipment will be available for continued long-term use. For comparison, fossil-fueled plants are retained in operation as, first, intermediate-load units and, finally, as reserve units. They are only decommissioned when continued maintenance becomes too expensive.

The economic incentives for refurbishing nuclear units are compelling. Even if it took $1 billion to refurbish the plant, the fact that the original costs have been fully amortized favors refurbishment over, for example, building a replacement coal plant. Besides, fueling costs for nuclear plants are, and will remain, less than half the fueling costs of coal plants.

The importance of this effort is now appreciated, and a program for extending plant life is under way. Necessarily, the requirements will vary from plant to plant, ranging from requalification of equipment to complex rework, including in some cases partial reconstruction of the plant. For some plants, running at reduced power would suffice. The variety of work to be done will undoubtedly keep the U.S. nuclear community busy over the next 20 years.

A forecast of the next period in U.S. nuclear power is possible based on the themes of introduction of new reactor designs and the extension of plant life. Nuclear power in the United States is scheduled to grow to about a 95,000-MW generating capacity by about 1995. Not all these plants will remain operational until 2010, but enough will be so that the occasional new nuclear plant in growth areas of the country should maintain a plateau of about 90,000 MW until installation of the new models begins. That should happen around 2010 to 2015, and this is the time I choose for the start of the second nuclear era.

Our pause in the installation of new nuclear power plants is being imitated elsewhere in the world, so we cannot expect export business to sustain our domestic nuclear industry indefinitely; instead, we must make innovative changes. Nuclear power is still an attractive option, but it will take ingenuity and perseverance, as well as an impeccable operating record over the next decade, for its potential to be realized.

AS SHOREHAM GOES[5]

"The Shoreham plant must open!" U.S. Energy Secretary John Herrington once declared. "If it doesn't, the signals will be the low point in this [nuclear] industry's history." By that criterion the signals are at their lowest point since the advent of the "peaceful atom," when the nuclear industry was conjuring up visions of electricity "too cheap to meter." The Shoreham Nuclear Power Station on Long Island will become the first completed nuclear facility in the United States to be abandoned before going into commercial operation. The settlement between New York State and the Long Island Lighting Company that brought this about was a stunning victory for grass-roots antinuke activists.

If it had gone into service, the Shoreham plant would have been a nuclear disaster waiting to happen. "No doubt about it, there is catastrophe ahead," George Henry, a former Lilco Shoreham inspector turned whistleblower, told Karl Grossman,

[5]Reprint of an editorial. Reprinted by permission from *The Nation*, V. 246:811–12. Je. 11, '88. Copyright © 1988 by The Nation Company, Inc.

author of *Power Crazy: Is LILCO Turning Shoreham Into America's Chernobyl?* (Grove Press). Another Shoreham inspector said that of the hundreds of thousands of mistakes and violations of standards built into the plant, fewer than half had been corrected. "You wouldn't believe what went on there," he said. "As God is my judge, if that thing is permitted to operate, a lot of people are going to die."

Because the plant is located only fifty miles east of New York City and on a densely populated island, an accident would have been catastrophic. The most recent Federal study of the consequences of severe accidents at every nuclear plant in the United States, done at Sandia National Laboratories in 1982, estimates that a core meltdown at Shoreham would cause 40,000 "early fatalities," 75,000 "early injuries," 35,000 "cancer deaths" and $157 billion in property damage. One expert on nuclear-plant-accident hazards, Richard Webb, regards that as a conservative projection, especially if the wind was blowing from the east, sending the plume of radioactive poisons from Shoreham directly into New York City. Then, he estimates, deaths would total "close to a million."

Not only potentially lethal, Shoreham is also the most expensive nuclear plant ever built—indeed, the most expensive power plant of any kind in terms of dollars per proposed kilowatt of electricity. Although cost overruns have been common in nuclear plant construction, Lilco's most recent estimate of $5.3 billion for Shoreham is 7,000 percent more than its original figures. When it first announced the facility, Lilco said it would cost "in the $65 million–$75 million range." Much of the excess is attributable to waste due to the power company's incompetence. The New York State Public Service Commission set a national record when it meted out a $1.4 billion penalty in 1985 to Lilco for pervasive mismanagement of Shoreham's construction.

Other utilities have engaged in heavy political lobbying but here, too, Lilco went to the extreme. On its home territory it has been the central corporate actor, so powerful, so busily manipulating the political process that predictions of "government by Lilco" gained currency on Long Island. Nevertheless, public resentment against the utility and opposition to Shoreham has been overwhelming: Polls show that nearly 80 percent of Long Islanders are against the plant and almost that many are willing to pay for it to be abandoned. In its Washington lobbying operation Lilco has enlisted former Federal officials such as Lyn Nofziger, who received nearly $1 million as a "governmental affairs consultant." A Lilco friend at the highest level of the Reagan Administration was Director of Central Intelligence William Casey, who, before he resigned his seat on the utility's board of directors, was a staunch promoter of Shoreham.

Some public-power advocates on Long Island protest that the settlement, negotiated by Gov. Mario Cuomo, does not go far enough. They wanted to eliminate not only Shoreham but the arrogant, inept utility itself. Taking its place would have been the publicly owned Long Island Power Authority, with directors elected by Long Islanders—and, analyses have shown, lower rates than Lilco's. The authority was created by the New York State Legislature and Cuomo in response to an intense grass-roots drive. The settlement that Cuomo opted for instead is generous to Lilco, whose stock immediately rose after the announcement. In return for selling the plant to the state for $1, the utility will be allowed annual 5 percent rate increases for a decade as well as taking generous tax write-offs. The public will thus pay for much of Shoreham's cost, including its dismantlement. The deal Cuomo struck was not the best, but consider the alternative: Shoreham in operation.

The closing of the facility provides a lesson to the rest of the nation: If people organize and fight back they can win, even against the formidable forces that backed Shoreham: Big Government (the White House, the Department of Energy and the Nuclear Regulatory Commission worked with Lilco), Big Business (the nuclear industry, which regarded Shoreham's operation as critical to its future, poured funds into the drive to save the plant), Big Banks (particularly the nation's largest, Citibank, which since the 1950s has functioned as a financial overseer of Lilco), Big Science (scientists from Brookhaven National Laboratory on Long Island have cooperated closely with the company in promoting Shoreham) and Big Media (*The New York Times* and *Newsday* have consistently demanded that the plant go on line).

The kind of dedicated, consistent, focused and massive grass-roots effort brought against Shoreham on Long Island, which stressed political action, civil disobedience and public education on nuclear power, can and should be duplicated elsewhere in the United States to prevent the opening of nuclear plants now under construction and to shut down the potential Chernobyls among us.

III. NATURAL GAS, COAL, HYDROELECTRICITY

EDITOR'S INTRODUCTION

Section Three of this volume considers three additional main-stays of the U. S. energy supply—natural gas, coal, and hydro-electric power. Unlike oil, national reserves of natural gas have not yet been extensively tapped, and natural gas has the addition-al attraction of being a clean-burning fuel that will not pollute the environment. A plentiful supply of coal also exists. Although the burning of coal produces sulfur and nitrogen oxides that pollute the environment and are considered a principal source of acid rain, industry has been developing new technologies that will re-duce these oxides dramatically. In the time ahead coal may once again become the workhorse of American energy production. Hydroelectric power has been an important generator of electric-ity in the past, but is not expected to expand significantly beyond its present capacity. Rather, it will continue to contribute a share of the national energy supply. If none of these sources is revolu-tionary, they yet offer the prospect of a stable energy source while a new energy technology is developed for the 21st century.

In the first article in this section, David Osborne, writing in the *Atlantic*, discusses the rich potential of natural gas for future energy. Osborne traces the history of natural gas in the last few decades, revealing that the presumption in the 1970s that we were running out of natural gas was mistaken, since vast reserves exist at a depth of four miles below ground. If the costs of explo-ration and production are kept below the price of OPEC oil, natu-ral gas would not only be a viable source of American energy but would also be abundant enough to "supply needs for a thousand years." In an article from *Maclean's*, John DeMont notes that many experts believe natural gas is about to make a comeback in the U. S. Reserves are also extremely plentiful in Canada which, with the signing of the Canada–U. S. Free Trade Agreement, may find a valuable market in California and other states. If OPEC raises the price of crude oil in 1990, the cheaper gas will become an attractive alternative.

The next two articles address the subject of coal. In an article from *Fortune*, Edward C. Baig discusses the new technology of fluidized-bed combustion that will remove sulfur oxides and substantially reduce nitrogen oxides as by-products of burning coal, thus improving upon and replacing the cumbersome and costly cleaning technology that uses "scrubbers." In a related article, Richard E. Balzhiser and Kurt E. Yeager, writing in *Scientific American*, point out the advantages not only of fluidized-bed combustion but also of coal gasification, which produces a fuel for boilers that burns cleanly and thereby eliminates the need for scrubbing. The two concluding articles in this section are concerned with hydroelectric power. Peter Kakela and his co-authors in *Environment* explain that "low-head" or short-fall dams can be a unique and valuable resource for local communities, if not on a national scale; and Sarah Cliffe in *Technology Review* notes that although many hydroelectric plants have fallen into disuse, they could make a comeback in the event of another energy crisis.

AMERICA'S PLENTIFUL ENERGY RESOURCE[1]

Ten years after the Arab oil embargo, the goal of "energy independence" remains as elusive as ever. The United States continues to import some 6 million barrels of oil a day, sending at least $60 billion abroad every year. Meanwhile, Iran vows to close the Strait of Hormuz—through which flows 35 percent of all oil traded in international markets—if Iraq's new Etendard jets attack Iranian oil installations.

As we wait helplessly for Iran's next move or OPEC's next embargo, however, the means to attain our goal lies hidden beneath our feet. Here in the lower forty-eight states, a resource exists that could sharply reduce our dependence on foreign oil within a matter of years. Unlike oil, this resource is still available in immense quantities. Unlike coal, it is so clean that we burn it in our kitchens. Unlike nuclear energy, it is safe and reliable. And

[1]Reprint of an article by David Osborne, freelance writer on political and economic affairs. Reprinted by permission of International Creative Management. Copyright © 1984 by David Osborne. Originally published in the March 1984 issue of *The Atlantic Monthly*.

unlike solar power, it is free of technological hurdles that might stand in the way of its exploitation.

This ideal resource is methane—in common parlance, natural gas. Methane is our quiet fuel: it heats more than half of American homes and powers more industry than oil or coal, yet it rarely makes news. For uses other than transportation, it is our most important fuel—and it could also power many automobiles if the shift had to be made. Italy and New Zealand boast several hundred thousand methane-powered automobiles; France, Egypt, Argentina, and Pakistan have initiated conversion programs; Canada is paying its citizens to switch, in the hope of seeing 2.5 million methane-powered cars by 1992. Here at home, hundreds of business fleets have been converted, and Ford is turning out cars designed to run on methane—which normally costs far less than gasoline at the pump.

If methane is so efficient and so economical, one might ask, why have we heard so little about it? The answer is depressingly simple: five years ago, many energy experts believed we were running out of natural gas.

This notion gained currency during the winters of 1976 and 1977, when pipelines to the Midwest and the Northeast ran short of gas. In the winter of 1977, one of the coldest in memory, a million people were thrown out of work, at least briefly, as gas-starved factories and schools were forced to close. Frightened by this calamity, and eager to sell Congress on the need for higher gas prices, Carter Administration officials insisted that by the end of the 1980s—*within thirteen years*—the nation's gas reservoirs would begin to run dry.

Today, the nation is experiencing a glut of gas so severe that producers are capping wells by the thousands. Respected authorities estimate that we have a seventy-five-year reserve, while visionaries speak in terms of hundreds of years.

Much of the gas was discovered during the drilling binge stimulated by the partial decontrol of gas prices in 1978. But even before decontrol, while the Carter Administration was still spreading its message of doom, drillers were hitting vast pockets of gas. The secret was depth: for years, petroleum geologists had assumed that no gas could exist below 20,000 feet (roughly four miles); today, the largest deposits are believed to be near or below that depth. "Any hole you drill in the deep Anadarko Basin [in Oklahoma] is going to produce some amount of natural gas," says

Robert Hefner III, the Oklahoma driller who proved the experts wrong. "Below about 15,000 feet, the sediments are saturated."

Most experts agree that the greater Anadarko Basin will produce at least 140 trillion cubic feet (tcf) of gas—beyond the 60 tcf or so (most of it shallow gas) it has already produced. At the consumption rate that prevailed before the recent recession—20 tcf per year—this translates into at least seven years' supply for the entire nation. And the Anadarko is only one of many deep basins, most of which lie virtually unexplored. Vast quantities of gas in shallower reservoirs also await development. In addition, both Canada and Mexico have reserves that may well be as large as ours; the United States could import far more from those countries than the one tcf per year it imports today.

And if all of these sources are not enough, immense reserves of "geopressured" gas are dissolved in hot, deep waters under Louisiana, Texas, and the Gulf of Mexico. If this gas can be produced at a price low enough to make it commercially feasible, it could conceivably supply all apparent needs for a thousand years.

To put all this in perspective, one must add that gas exploration is still in its infancy. Until 1978, when the gradual decontrol of new-gas prices was initiated, most of our gas had been discovered inadvertently, during the search for oil. Except in a handful of states, the industry did not actively look for gas before five years ago. Spurred on by prices that have risen sharply in that period, drillers have since sunk wells everywhere from Oregon to New York—and have found gas. Oil does not normally survive the fiery temperatures found below 15,000 feet; hence a century of shallow drilling may have already drained a good deal of our oil reserves, at least in the lower forty-eight states. But we have just scratched the surface of our potential gas supplies.

Geologists have known this for several years. In 1981, the presidents of the American Gas Association and the Gas Research Institute, along with Robert Hefner and other drillers, said as much to a congressional subcommittee. They also described what we should do next: increase our production of gas; develop technologies to tap the geopressured gold mine under the Gulf Coast; substitute gas for some of our midwestern coal, to cut down on acid rain; and start converting our auto fleet to methane. By raising our natural-gas consumption by just 33 percent, we could replace 2 million barrels of imported oil per day, keeping $20 billion a year from going abroad, creating thousands of jobs, and pouring perhaps $6 billion in taxes and royalties into the federal treasury.

Nothing is ever quite that simple, of course. And conversion to natural gas is proving to be the opposite of simple. Since 1981, domestic gas has *lost* ground to foreign oil. During our first natural-gas glut in decades, consumption has plummeted and prices have skyrocketed, thwarting the laws of supply and demand. Cheap gas has been shut in the ground while expensive gas has been produced. The cost of heating with gas has doubled in just four years. The natural-gas market, it appears, has lost all semblance of rationality.

Congress, bewildered by this upside-down world, is not sure where to turn. In the confusion, the Reagan Administration has seized the initiative, proposing to decontrol *all* gas prices and return gas to a presumably free market. As so often happens, crisis has triggered not innovative thinking but conventional wisdom, bringing forth strategies developed to meet yesterday's problems. In this case, the proposed solution is not merely irrelevant; it will probably make matters worse.

To understand how we could go from shortage to glut in the space of a few years, we must understand the effect of federal price controls on the natural-gas industry.

Interstate gas pipelines were placed under federal regulation in 1938, soon after they began snaking their way across the country. Like utilities, pipelines are what economists call "natural monopolies." Since it would be wasteful to build three or four pipelines to serve the same route (as it would be wasteful for three or four gas utilities, each with its own labyrinth of underground pipes, to serve the same customers), most areas are supplied by one or, at most, two. Thus, once a utility signs up with a pipeline, it is stuck; if prices for gas from that pipeline are raised, the utility has nowhere to turn for cheaper supplies. Similarly, when the utility raises its rates, the individual consumer is powerless. Free from competition, both utilities and pipelines could impose huge rate increases on their customers if the prices they charge were not regulated.

In 1954, when the state of Wisconsin sued Phillips Petroleum for price-fixing, the United States Supreme Court extended federal price controls from pipelines to producers of gas. It based its decision largely on investigations by the Federal Power Commission, which had found creeping monopolization in the gas fields as the major oil companies came to dominate the industry. In any given field, according to FPC data, only one, two, or three companies produced gas, and only one pipeline bought it—hardly an indication of the thriving competition upon which a free-enterprise system depends to keep prices in line.

The oil companies have fought price controls since the late forties, when they were first proposed in Congress. Twice during the fifties the industry pushed decontrol bills through Congress, only to see them vetoed—once by President Truman, once by President Eisenhower. During the sixties, the struggle moved back to the courts. In 1968, the Supreme Court ruled, as it had in 1954, that the intent of the Natural Gas Act justified price controls on producers as well as pipelines, once more citing the absence of true competition in the gas fields.

Because the 1938 act applied only to pipelines in *interstate* commerce, however, price controls were never extended to gas that did not cross state lines. In 1973, when OPEC quadrupled the price of its oil, this unregulated *intrastate* gas started to follow oil up the price ladder. The reason is simple: in industrial markets, which account for more than 50 percent of all gas consumption, gas competes with oil. As long as industries that can just as easily use gas must pay more for oil, natural gas obviously will be in great demand. Where gas prices are not controlled, rising demand will pull them up—as it did in Lousiana, Texas, and Oklahoma, the producing states with the biggest internal markets.

By 1975, intrastate prices were up to $1.40 per thousand cubic feet (mcf), two and a half times the interstate price. Being intelligent businessmen, gas producers were more interested in selling their gas for $1.40 than for 57 cents. When they could not get the higher price, they often chose not to sell at all, preferring to await better days. Soon new interstate sales began to dwindle; pipelines had already begun curtailing sales to industrial users. By 1977, the shortage was severe enough to close schools and factories. Meanwhile, Louisiana and Texas were floating in gas.

Anyone familiar with the gas industry understood that the shortage was artificial, a result of the unwillingness of gas producers to sell at controlled prices. Less astute observers looked at the industry's figures for "proven" reserves, which were steadily falling, and concluded that the nation was running out of gas.

Ever since the 1968 Supreme Court decision, the industry had warned that the continued imposition of price controls would limit gas production and create a shortage. Sure enough, as early as 1969, figures for annual reserves began to decline, a trend that was to last for a decade. Suspicious of a possible connection, the Senate Antitrust and Monopoly Subcommittee asked the Federal Trade Commission to investigate the reliability of these industry figures. Most companies whose books were subpoenaed refused to cooperate. The books of those who agreed, however, clearly revealed manipulation of reserves figures: one set of numbers, the highest ones, went to the banks to facilitate loans; a middle set went to the Internal Revenue Service; and the lowest figures went to the American Gas Association, which published the national totals.

As the shortage of gas available for interstate sale became more acute, however, few among the public or the press were aware of these manipulations. The average person likewise failed to understand the limited importance of the industry's declining figures, accurate or not. Proven reserves—the estimated amount of gas that can be pumped out of the ground from proven reservoirs at existing prices—are nothing more than inventory. When newsprint inventories fall, the paper industry does not declare that we are running out of trees. But when gas inventories fell, the oil industry announced that the nation was running out of gas. In 1974, Mobil and Shell warned that we had enough gas left for only twenty to thirty years; in 1976, Exxon's estimate was fifteen to twenty years. Meanwhile, United States Geological Survey (USGS) estimates ranged from forty-four to a hundred years. Even the CIA got into the act, concluding, in 1977, that we had fifty to sixty years' supply. The perception of shortage did not, of course, damage the case the oil companies were busy making before Congress: that if they were to get every last drop out of the ground, prices had to be decontrolled.

To most people, a shortage was a shortage. President Carter was no exception. Dismissing as absurd the deep-gas drillers who came to Washington with tales of plenty, the Carter Administration bought the oil companies' line: that most of the cheap, shallow gas had already been found; that gas did not generally exist below 15,000 or 20,000 feet; and that future drilling would yield ever less at ever higher prices. But even as Carter was preparing his major energy address (which became known as the Chicken Little speech, for its "the sky is falling" tone), government scientists were gathering data that contradicted it.

Initial reports from deep wells in the Anadarko Basin and the Rocky Mountain Overthrust Belt indicated enormous reserves of gas. And when Vincent McKelvey, then the director of the USGS, examined data on the geopressured zone under the Gulf Coast, he discovered "almost incomprehensibly large" deposits of natural gas. The geopressured reserves, even by low estimates, he announced in a 1977 speech, amounted to "about ten times the energy value of all oil, gas, and coal reserves in the U.S. combined." No one knew whether the geopressured gas could be economically produced, however. When Dr. Christian Knudsen, a former Exxon scientist who joined the Federal Energy Administration in 1974, ran a reserves study for the Department of Energy, he ignored the geopressured reserves. Playing down the potential of deep gas, as well, Knudsen nevertheless concluded that the United States held enough gas to last us well into the next century.

The good news was not welcome at the Department of Energy, for it contradicted the reasoning behind the entire Carter energy program, built as that program was on conservation, higher prices, and synthetic fuels. When Knudsen's conclusions became known, a consultant from Penn. State University was called in to double-check his data. To the chagrin of Knudsen's superiors, the consultant concluded that Knudsen was being too conservative. Frustrated, the DOE next brought in an expert from Shell Oil, who did suggest that some figures in the report should be significantly lower, to conform with a supply curve from the Stanford Research Institute—a curve whose assumptions Knudsen was never allowed to examine. Finally, a week after Knudsen had presented his final report, he was called into Deputy Secretary John O'Leary's office for a dressing-down. Over the weekend preceding Carter's energy speech, the report was frantically rewritten to support the Administration's position. Knudsen, meanwhile, was fired.

Vincent McKelvey soon met a similar fate. After two speeches describing natural-gas reserves in terms that far surpassed those used by the DOE, McKelvey was asked for his resignation. To this day, he has been given no explanation. Those around him, however, never doubted that his position on gas reserves cost him his job.

Knudsen and McKelvey were not the only people talking about plentiful supplies. Executives from gas-producing companies also came to Washington to spread the good news about gas. Rather than investigate their reports, however, President Carter turned down the thermostat, put on a sweater, and pushed several important bills through Congress.

Two dealt with natural gas. The Powerplant and Industrial Fuel Use Act, an attempt to conserve our supplies of fuel for home heating, sharply restricted the use of gas by utilities and industry. The Natural Gas Policy Act (NGPA), an effort to encourage new production, gradually raised prices for all gas from wells begun after 1977, and mandated final decontrol of this "new" gas in 1985. Carter's bill also eliminated the dual market for gas, imposing interstate prices on intrastate gas. And, largely through the efforts of Robert Hefner, Congress immediately decontrolled gas extracted from depths greater than 15,000 feet.

Unfortunately, the two bills had contradictory results. The NGPA brought in gushers of new gas, while the Fuel Use Act helped constrict the market for it.

The higher prices allowed under the NGPA immediately brought forth a surge of new gas, proving that the shortage had been an artificial one. The prospect of rising profits also brought on the biggest drilling binge in the industry's history. Stirred by visions of OPEC oil hitting $100 a barrel, speculators armed with bank loans sank hundreds of wells in the Anadarko and the Rocky Mountain Overthrust Belt. One vast field after another turned up in the deep basins, and significant new supplies were found even in shallow fields. Normally, wildcatters consider themselves lucky if one out of ten wells produces; during the late seventies, *seven* out of ten deep wells in the Anadarko hit gas. This one basin produced 6 trillion cubic feet of new gas in four years, or roughly one twelfth of total U.S. consumption during the period. Total additions to reserves in the lower forty-eight states increased from an average of less than 10 tcf during the seventies to more than 20 in 1981, as the decontrolled gas became available.

Why the sudden discovery of gushers below 15,000 feet, after years of declining reserves? The obvious answer is that before deep gas was deregulated, very few wells had been drilled below 15,000 feet. More to the point, once the depths were probed, they turned out to be literally saturated with gas.

The geological explanation for this auspicious circumstance exhibits the pleasing simplicity of science. At 12,000 to 15,000 feet in most sedimentary basins, the weight of the overlying rock becomes so great that all porosity in the sediments is squeezed shut, creating a relatively impermeable layer. For many years, geologists assumed that beneath this level porosity remained close to zero, except in very unusual circumstances. The deep drilling of the past decade has found just the opposite, however: below the impermeable layer, porosity suddenly increases as the sediments become saturated with gas under great pressure—the so-called "geopressured" gas.

The impermeable layer acts as a seal. As gas accumulates, it cannot move toward the surface, as it would in shallower regions. Only when an enormous amount of gas has built up, creating pressure great enough to fracture the rocks, can it burst through toward the surface. As a result, since gas has been accumulating over eons, the geopressured zone is likely to be full today.

Above 12,000 to 15,000 feet, the situation is quite different. Accumulating gas will always escape to the surface unless it is trapped by a "cap rock"—a nonporous stratum of rock whose shape allows gas to collect beneath it. Cap rocks are present only in a tiny fraction of the upper domain, however. Consequently, most gas originating above the geopressured zone eventually escapes, venting into the atmosphere or the oceans.

Dr. Paul H. Jones, who spent thirty-three years with the USGS, has actually mapped the impermeable layer in the Gulf Coast area, using data from thousands of wells to show the depth at which porosity suddenly increases. Many geologists, however, still believe that porosity must decrease the deeper one drills. When it increases, they label it a "pressure anomaly," a local exception. But the facts now suggest otherwise, and conventional opinion is shifting as the industry gains experience drilling for deep gas.

Perhaps the nation's most authoritative source on gas supplies is the Potential Gas Committee, whose members represent all branches of the industry, as well as the government and academia. As of the end of 1982, the committee estimated, the United States held 875 tcf of conventional undiscovered gas, plus roughly 200 tcf of proven reserves. Conservative estimates of the amount producible from tight-sand formations, coal seams, and other "unconventional" resources (excluding geopressured gas) are in the 400-to-500-tcf range, bringing total reserves and resources to between 1,500 and 1,600 tcf—a seventy-five-year supply at normal rates of consumption. Comparable resources may exist in Canada and Mexico. A few experts continue to dispute these figures, though most of the low-end estimates (in the 250-to-400-tcf range for conventional resources) were discredited by the surge of discoveries in the early eighties. The low estimates have two defects in common: they assume that gas and oil are similar, and they base projections for new discoveries on past trends. In reality, almost all oil is found above 15,000 feet and, as we now know, a great deal of gas is found below 15,000 feet. Drillers have been combing the country for oil for eighty years, whereas they have been actively searching for gas—and drilling deep—for only five. Predicting gas discoveries on the basis of past trends makes as much sense as predicting solar-energy production from the historical record.

Even the latest estimate by the Congressional Office of Technology Assessment perpetuates the errors. The OTA takes the path of least resistance, splitting the difference between low and high estimates with a range of 400 to 900 tcf, or twenty to forty-five years' supply. (The OTA figures are for conventional gas in the lower forty-eight states; thus they exclude hundreds of trillions of cubic feet in tight-sand formations, in coal seams, and in Alaska.) Neither the importance of deep gas nor the fledgling nature of gas exploration appears to be understood by the OTA authors. In addition, their predictions are based on "present and easily foreseeable technological and economic conditions." Presumably, technology will improve and prices will rise. To their credit, the authors tacitly admit these problems:

Furthermore, some scientists believe that those estimates that invoke current technology and economic relationships—the great majority—are simply irrelevant, whether or not they are correct within the constraints of these assumptions. These scientists believe that both the inexorable advance of technology and rising prices that reflect resource scarcity will constantly push outwards the boundaries of the recoverable resource base. As noted previously, the history of resource estimation in general tends to support this view; cycles of predictions of scarcity followed by radical upward revisions in resource assessments appear to be common for nonrenewable resources.

By this reasoning—which is little more than common sense—even the Potential Gas Committee's estimates will grow over time.

Guessing what is hidden deep within the earth will always be problematic. To temper our instincts for caution, however, we would be wise to look at the history of oil-supply predictions. In 1908, the USGS estimated maximum future U.S. supply at 23 billion barrels; in 1914, the Bureau of Mines dropped that figure to 6 billion. Today, more than 100 billion barrels have been produced. In 1920, the USGS director warned that U.S. production had almost reached its peak, sparking a decade of interest in synthetic-fuel production. The Department of the Interior sounded the alarm in 1931, 1939, and 1949. Yet by 1949 industry experts had far more experience as oil prospectors than their counterparts today have as gas prospectors.

Fifty years ago, few people imagined that we might someday drill for oil offshore or in Alaska. Today, analogous advances seem perfectly plausible. Consider the task of sinking a drill five miles into the earth, where pressure changes so suddenly and dramatically that wells are easily ruined. Ed Schmidt, an energy consultant who knows a great deal about the Anadarko Basin, compares deep-drilling technology today with aviation technology in the midthirties. The Boeing 747, he points out, is one hundred times as productive as the DC-3.

The gas resource that is most dependent on technological innovation is geopressured methane. None of the figures cited above include geopressured reserves, which most estimates put at between 3,00 and 100,00 tcf—or 150 to 5,000 years' supply. Unfortunately, however, geopressured gas is dissolved in hot salt water at relatively low densities, deep underground. To produce it, drillers must sink wells 12,000 feet or more, separate the gas and the water, and dispose of the water—perhaps by pumping it back down exhausted wells. None of this would be cheap. But even if only a fraction of this gas could be produced at market rates, it would change our supply picture dramatically.

Paul Jones, the nation's foremost expert (and most optimistic visionary) on geopressured gas, believes that the Gulf Coast region holds 50,000 tcf in the geopressured zone above a depth of 25,000 feet, plus more at deeper levels. He thinks that a great deal of it can be tapped for commercial use. The numbers lead him to the conclusion that "we have more recoverable natural gas in the world than we will ever be able to use." "All of the petroliferous basins of the world are geopressured, and all of the geopressured waters are saturated in natural gas," Jones says. "The amount in the Gulf Basin alone staggers the mind."

During the late seventies, the government drilled a number of test wells in the geopressured zone. After encountering a series of technical difficulties, DOE officials concluded that the gas could not profitably be tapped at prevailing prices. But, according to Jones, the oilmen hired to run the tests used oil-well technology when they should have drilled water wells. Jones has patented a drilling process that, he says, will produce plenty of gas at market rates; it involves drilling deep enough to tap superheated water that will blow off as steam, releasing enough pressure to allow the gas to come out of solution in the reservoir. Since Jones has not been able to raise the capital necessary to test his ideas, however, no one knows whether they will work.

"Gas from the geopressured zone is by far the cheapest gas the country can produce," Jones says. "The deeper you go, the larger the gas reservoirs in terms of recoverable gas." Jones believes this is largely because gas grows dense under great pressure and expands considerably when brought to the surface. One deep well in Louisiana, Jones points out, produced 3.2 billion cubic feet (more than $5 million worth of gas) in just twenty-nine days— roughly twenty-five times the flow rate of a good shallow well. On the basis of photographs of steam billowing out as the well first came in, Jones is convinced that the drill reached the geopressured zone, blowing off enough pressure to let the gas come out of solution deep underground.

For some reason—perhaps because we will not need geopressured gas for decades—the industry has largely ignored its potential. Yet most industry spokesmen assume that advancing technology and higher prices will eventually allow us to manufacture synthetic gas from coal. Separating gas from hot salt water ought to be much simpler.

Industry leaders such as Henry Linden, president of the Gas Research Institute, and George Lawrence, president of the American Gas Association, believe we could increase our gas production 50 percent by the year 2000. Such opinions are rare among academic energy experts, however. Foundation and academic studies of future energy supplies invariably tip their hats to natural gas, and then ignore it. In part, this may be because the authors of such studies have generally been led into the subject by the oil crisis. "People who have been studying the oil markets tend not to be the same people who have studied the gas markets," explains Harry G. Broadman, of Resources for the Future, Inc., which administered a 1979 Ford Foundation study. "We're only now beginning to bridge the gap."

Many experts also have a decade-long investment in the argument for conservation, which could be undermined by tales of huge energy supplies. (In reality, both oil and gas are so expensive today that conservation is cheaper than any new source of energy, and should be encouraged.) As long as low estimates like those in the OTA report are prevalent, these experts won't have to take a closer look.

The contrast between energy-study teams at the Harvard Business School and at Georgetown University's Center for Strategic and International Studies indicates the range of current thinking in academia. In 1979, a Harvard study called *Energy Future* estimated that current gas-consumption levels could be maintained for only twenty-five to thirty years before sharply higher prices would have to take effect. In 1982, one of the authors of the study, Daniel Yergin, called the United States "an aging producer" of gas that was "outrunning its geological base." Like the Harvard study, a CSIS report briefly mentioned optimism among some gas producers but generally ignored gas as a potential solution to our dependence on foreign oil.

Today, Yergin remains cautious about gas supplies, though he agrees that "there is a high degree of uncertainty, and people can make good cases on either side." Attitudes changed at Georgetown when Henry Schuler, once an oil-company executive, joined the staff. "My thinking is that natural gas is—bar none—the most attractive alternative to foreign oil," Schuler says. "The resource base is enormous in this country, and it's available in the lower forty-eight states. Obviously natural gas is environmentally the most attractive fuel of any, because it burns so cleanly. The transportation system is in place for natural gas, so you don't have the problems of coal, with the railroads in the East being really unable to carry a lot more coal. The technology is in place. And there's no doubt in my mind that in a deregulated market natural gas can compete on a price basis with imported oil. We should be optimizing both the production and the consumption of gas."

One concern both Yergin and Schuler discuss is the price at which gas can be produced. With the average wellhead price having tripled since 1978, the obstacle to the optimal use of natural gas is no longer short supplies but high prices. In some parts of the country, gas already costs more than residual fuel oil. (The cheapest oil available to industry, residual oil is the heavy residue left after crude is refined.) According to the American Gas Association, 52 percent of electric utilities and industrial consumers—the largest market for gas—have the capacity to use either fuel. Thus, in the short term it hardly matters whether we can produce more gas; if it costs more than oil, no one will want it.

The average wellhead price of gas was $2.63 per thousand cubic feet last October—roughly $1.45 for old interstate gas, $3.55 for new. None of these prices include processing and transportation costs, which amount to another $1.20 to $1.50. Pipelines depending upon large quantities of new gas were, of course, paying more than $2.63. At $2.85, gas becomes as expensive to use as residual oil in some markets—particularly those where industrial users can burn the least expensive high-sulfur oil without worrying about environmental problems. The break-even point is generally between $2.85 and $3.50, though it goes higher in a few markets. As a consequence, the price of gas is already close to parity with that of its most competitive fuel, residual oil.

The deeper a driller has to go for new gas, the more expensive it becomes. Hence, a key question in the natural-gas debate is whether large supplies can continue to be produced at prices below those of foreign oil. Heavy industry and electric utilities, which buy most of our gas, remain uncertain about the answer.

Gas *should* be cheaper than foreign oil, in a rational marketplace. Today, 10 to 20 percent more gas is available than the market can absorb; thus, prices at or below those of oil obviously have been adequate to stimulate exploration and production. And current gas prices are *higher* than they need to be. The 1979 decontrol of gas found below 15,000 feet created a sudden price incentive to drill that deep, more than three times the depth of the average well sunk before 1979. As a result, a great deal of the gas between 5,000 and 15,000 feet—which would be cheaper to produce than deep gas—was leapfrogged. By 1983, expensive deep gas (and gas from other unconventional sources also decontrolled in 1979) represented 8 or 9 percent of all gas in interstate pipelines but accounted for 23 percent of the total cost for interstate gas. When all new gas is decontrolled next year, drilling will return to the shallower, cheaper reservoirs, lowering the average costs of drilling.

Most gas is much cheaper to produce than oil, and always will be. Oil must be pumped out of the ground and refined, while gas rises on its own and must simply be stripped of moisture and contaminants, such as sulfur. In 1975, the last time the Federal Power Commission tried to estimate the total exploration, production, and development costs for gas, it came up with a figure of 67 to 80 cents per thousand cubic feet, which allowed for a 15 percent rate of return on investment. (This estimate primarily reflected the cost of shallow gas, since the production of deep gas was rare in 1975.) Today inflation has probably raised the figure to $1.50.

In 1983, the federal Energy Information Administration made a similar study. In a working draft, the EIA concluded that the vast majority of new gas in the nation could be produced for under $3.00 per mcf, allowing for a 15 percent profit. Even gas from most fields at 17,500 feet, the deepest the EIA analyzed, fell under the $3.00 mark. The study did not include exploration costs, but this omission was offset by the fact that its base year was 1981. Since then, as the gold-rush fever in the gas fields has subsided, drilling costs have fallen 30 percent, interest rates have dropped even further, and federal taxes have been cut sharply.

The cost of deep gas is a subject of controversy in the industry today. Looking at the 1979–1982 period, when most deep drilling was done, many in the industry conclude that deep gas can only rarely be produced for under $5 per mcf. Others point out, however, that the 1979–1982 era was marked by extreme inflation in the industry, as demand for drilling rigs and supplies intensified. They also note that most of those drilling deep were new to the game, lacking experience with the tricky geology one finds below 15,000 feet. As a result, mechanical errors were common and many promising wells were ruined.

Bob Hefner, whose GHK Companies drilled the first deep well, in 1969, probably knows more than any other man alive about both deep drilling and the Anadarko Basin. Hefner is known in the industry as a gambler and an optimist, and some people discount much of what he says. But some of his maverick notions, particularly about the existence and scale of deep-gas reservoirs, have proved correct in the past. And his company has as much experience drilling for deep gas as any other. Today, Hefner believes that most deep wells will be quite profitable at $3.00 to $4.00 per mcf. Others in the industry agree, including

Brian Whitworth, a vice-president at Phillips Petroleum, which has done substantial deep drilling in the Anadarko.

Hefner does not deny that costs rise geometrically as one drills deeper—from $200,000 for the average 4,000-foot well to between $5 and $10 million for a well of more than 20,000 feet. At great depths, one encounters sudden changes in pressure and rock density, which require quick adjustments in drilling speed, drill weight, and the type of drilling muds used to lubricate the bit. However, Hefner says, thanks to economies bred of experience his company today is drilling significantly faster than it was just a few years ago, and twice as cheaply. His people have learned so much about the geology of the deep Anadarko that they can now anticipate where they will hit sudden changes in pressure and density. They also know more about where to drill for the biggest payoffs.

Perhaps the greatest advances have been in drilling technology, however. Bits and bearings made of new, tougher alloys have been introduced. Rather than turning three or four miles of drill string from the top, motors are now used at the bottom of the hole as well. And drilling muds that stand up to the fierce heat and pressure found at 15,000 or 20,000 feet have been developed. The water-based muds used in the past often cemented up under intense heat and pressure, blocking completion of the well. Today, oil-based muds are used. Other technologies—such as "continuous logging" to let the driller know immediately what pressure and temperature and geological formation one bit is encountering—are still being perfected.

The other factor often ignored by those lacking experience in deep drilling is the very nature of deep gas. Unlike oil or shallow gas, gas below 15,000 feet seems to be pervasive in sedimentary basins. Trapped by the impermeable layer, it saturates huge areas—greatly reducing the driller's chances of hitting a dry hole. In addition, deep gas is compressed, so volumes of production, like drilling costs, increase radically with depth. During the recent deep-gas bonanza, wells that will produce for years were pouring forth enough gas to repay drilling expenses in a matter of months.

The upshot of all this is that the current wellhead price for new gas should be more than enough to pay the costs of exploration and production for quite a few years. As fields above 15,000 feet are played out in coming decades, more deep gas will have to be found, and prices will gradually rise. The higher demand created as we replace foreign oil will also increase prices. But since pipeline costs do not increase as more gas moves through the line, transmission costs per unit of gas will *fall*. And, given the extent of conservation efforts since the first oil shock, demand should not escalate very rapidly. The average home now uses 20 percent less energy than it did ten years ago, and industry has probably done even better. Conservation efforts are still in high gear, which means that total energy consumption—which has fallen ten percent since 1973—should rise very slowly in the future, if at all.

One must also remember that oil prices are bound to rise. Domestic oil is much scarcer than gas, and as we go offshore and to Alaska for a greater proportion of it, costs will escalate. Foreign-oil prices, of course, are still influenced—if not controlled—by a cartel.

For all these reasons, gas should be cheaper than oil for as long as we use either fuel. Today, it should be much cheaper than OPEC oil, everywhere in the country. In a conventional market-place, the current glut would have drastically lowered prices, stimulating demand and replacing a great deal of foreign oil. Instead, average wellhead prices rose until February, 1983, and then dropped by only a few cents. As a consequence, gas has lost ground to foreign oil in the battle for market share.

Industry spokesmen cavalierly blame this problem on continued price controls—as if ceilings somehow prevent prices from falling. The real story is very different, however. Gas prices rose despite sinking demand because buyers were panicked by tales of long-term shortages. They have remained high because the absence of genuine price competition has allowed major producers virtually to ignore the laws of supply and demand.

During the late seventies, fear of another shortage was so widespread that it triggered a wild scramble by gas-pipeline companies for guaranteed supplies. With old gas still under price controls, the major oil companies, which own most of the old gas, had little reason to increase their production from old fields. Some oilmen have even admitted that significant amounts of old gas were held out of production so that greater quantities of the more profitable new gas could be sold.

Faced with limited supplies of old gas, pipelines and utilities were forced to consider more-expensive options: new gas, deep gas, and foreign gas. In the mad rush to line up supplies, pipelines bid up prices for deep gas to $9 and $10 per mcf—four times the price of new shallow gas. Most agreed to "take-or-pay" contracts, which obligated buyers to pay for most or all of a producer's gas—whether they needed it or not—for five, ten, or even twenty years.

New shallow gas, permitted to rise gradually in price until full decontrol in 1985, was also in great demand. Most contracts called for the gradual price escalations allowed by law and sharp increases upon decontrol. Thus, a producer might be guaranteed $5 per mcf in 1985, even if the market price is only $3.

The price charged by Canadian suppliers, who were not subject to American controls, hit $4.94. Some pipeline companies even signed long-term contracts for more-expensive liquefied natural gas from overseas. Driven by the expectation of shortage and allowed by regulations to pass rising costs on to their customers, pipelines had every incentive to nail down long-term supplies and no incentive to hold the line on price. This madness built huge price increases into the system, which hit consumers hard as the new gas came into use in the eighties.

As prices rose, demand fell. Industry had initially been scared off gas by the shortages of the mid-seventies, during which many companies had been forced to switch to fuel oil. In 1979, the Carter Administration's Fuel Use Act institutionalized the switch. It held gas consumption by utilities at the low levels established during the shortages, prohibited extensions of many gas contracts, banned construction of new gas and oil boilers, and ordered most utilities to shift from gas to coal by 1990. The act also restricted industrial use of gas. Until the Iranian-inspired oil shortage of 1979, Congress and the President were actually moving to force electric utilities to switch to foreign oil.

The day the Fuel Use Act was passed, many people in Carter's Department of Energy understood its flaws. "By the time it was signed, we knew there was plenty of gas around," says David J. Bardin, who had to administer the act. When the 1979 oil shortage hit, Bardin began granting any and all exemptions allowable under the bill, urging industry to shift back from oil to gas. And in 1981, the Reagan Administration persuaded Congress to knock out the act's most restrictive provisions. By then, however, gas prices had escalated so wildly that oil was actually cheaper in many parts of the country.

When the recession hit, in 1982, a full-scale glut of gas was on hand. Drillers who had borrowed millions to sink deep wells, expecting a return of $9.00 per mcf, saw prices plunge to the $3.00 range. Chaos hit the industry, as boom turned to bust. Wells were capped, bankruptcies were filed, and several banks with bulging gas portfolios collapsed.

Meanwhile, pipelines found themselves stuck with more gas than they needed, on take-or-pay contracts. A survey by the Interstate Natural Gas Association of America projected that pipelines would pay $3.3 billion in 1983 for gas not taken. Consumer outrage grew as pipelines cut off their cheap suppliers and continued pumping the expensive gas with which they had so recently saddled themselves. In just one example, Columbia Gas Transmission Corporation, which supplies much of the mid-Atlantic region, stopped buying $1 gas from shallow wells in Ohio, West Virginia, and Pennsylvania while continuing to pay $9 for gas from deep wells in Louisiana and Texas.

Unable to sell all this expensive gas in a recessionary market, some pipelines and utilities threatened to follow the drillers into bankruptcy. Many American manufacturers dependent upon gas teetered on the same brink. Misguided notions of shortage had finally collided with the reality of abundance, shaking the industry to its roots.

The only way to bring the gas industry to its senses, according to the oil companies that dominate it, is to return it to the free market. To do so, the Reagan Administration has proposed to decontrol all gas, old and new, repeal the remaining provisions of the Fuel Use Act, and allow all long-term contracts to be broken in 1985. Suddenly the nation's entire supply of gas would be up for sale at whatever price buyers and sellers could negotiate. Let prices find their own levels, the Administration reasons, and everything else will fall into place.

This argument would make sense in a market where many sellers competed, each trying to attract buyers by offering the lowest price. But genuine competition does not exist in the natural-gas industry.

Oil and gas executives have argued for decades that the competitive nature of their industry is obvious, since thousands of individual companies produce gas. The rule of thumb, they point out, is that eight or fewer firms must control at least 50 percent of production in an industry before anticompetitive pricing is possible. In natural-gas production, sixteen companies are needed to hit the 50 percent mark.

Unlike autos or appliances, however, gas is normally sold through long-term contracts, guaranteeing supplies for ten or twenty years. Thus, to measure price competition at any point in time, one must count the firms that control all gas currently available for sale, not all gas currently under contract. In 1973, the last time the federal government investigated ownership of these "uncommitted reserves" in most producing regions, the top eight firms—almost all of them major oil companies—controlled well over 75 percent. Of the top four producers in each of the six regions studied, all but two were major oil companies. The exceptions were a gas-pipeline firm and a large independent gas producer.

The Federal Power Commission staff found a similar pattern when it investigated offshore and onshore reserves. Small companies could not afford to bid for offshore leases. Onshore, the independents normally drilled on land owned by one of the majors, usually selling their discoveries to the company that owned the land.

The major oil companies not only owned most uncommitted reserves but also bid jointly for many leases, both onshore and offshore. David Schwartz, who was the assistant chief in the FPC Office of Economics when the study was done, describes the process: "Exxon, Mobil, and Texaco acquire acreage and set up an operator on a lease, who markets for all the owners. So each stream of gas is sold not competitively but together. It's as if U.S. Steel and Bethlehem Steel set up a joint plant—we'd never accept it in any other industrial sector. I contend that it is anticompetitive. By marketing through these joint ventures you obviously peel off the possibility of the majors competing against one another."

Some economists point to a countervailing force: the fact that only one pipeline exists to buy gas in many areas. This gives the pipeline "monopsony power," or the ability to demand low prices because, inasmuch as many areas are covered by only one pipeline, producers have nowhere else to take their gas. Unfortunately, the situation is not that simple. Because it is difficult to store large amounts of gas for periods of shortage (or to buy new gas on a spot market, since expensive pipelines must be built to hook up new fields), pipelines must assure themselves of long-term, uninterrupted supplies. For this reason, interstate contracts have historically been for twenty years or more, though the shortages of the 1970s brought the average down, and it is perhaps ten years today. In many areas, the only companies that can provide enough gas are the majors—which, as we have seen, often sell as one. Having invested heavily to lay their gathering lines, the pipelines cannot afford to turn away from the majors' gas. Thus their monopsony power is offset by the major producers' monopoly power.

Federal regulations that allow pipelines automatically to pass the cost of their gas on to the utilities—which are in turn allowed by state utility commissions automatically to pass it on to their customers—further destroy the incentive to bargain for lower prices. Given this situation, pipelines have usually bargained for supplies, not for price. Long ago, they accepted federal ceilings as the prices they paid the majors, though they have been perfectly free to demand lower rates.

To compound the problem, all thirty-one of the major interstate gas pipelines also produce gas, through subsidiaries or affiliated companies. Some buy as much as 35 percent of their supplies from themselves; the average is 12 percent. According to David Schwartz, "None of the thirty-one has acquired any acreage independently. They produce in combination with the majors." Schwartz believes that this creates a community of interest between pipelines and producers, who are supposedly bargaining "at arm's length."

In addition, federal regulations allow pipelines to pay their own producing affiliates the same price they pay others. As everyone in the industry acknowledges, this hardly encourages pipelines to seek lower prices, since lower prices would cut their profits. Just last year a judge at the Federal Energy Regulatory Commission (the new name for the FPC) found Columbia Gas Transmission guilty of cutting back purchases of low-cost gas from other producers in order to continue buying its own high-cost gas, in quantities larger than those specified by its contracts.

Many oil companies also own gas pipelines, particularly intrastate pipelines. Pipelines and producers are further linked through interlocking boards of directors. During the early seventies, as the shortage of interstate gas developed, the FPC commissioners allowed any interstate pipeline whose deliveries had been curtailed to pay higher prices temporarily, to secure additional supplies. When the FPC staff investigated these emergency sales, they found some eighty cases in which the same person sat on the board of both the buyer and the seller.

Historically, all of these forces have combined to let large producers virtually dictate their terms. The typical results have included "most-favored-nation" clauses, which automatically raise producers' prices to the highest figure secured by any other producer in the area; "indefinite price escalators," which automatically increase prices whenever federal ceilings rise; and take-or-pay contracts.

Some observers argue that though this analysis was valid in 1973, aggressive expansion by independent drillers has changed the picture. Because the independents owned more than 70 percent of all new-gas reserves by 1982, these observers point out, the share of production controlled by the twenty largest producers fell from 61 percent in 1975 to 47 percent in 1981. But without knowing how much of the independents' gas is marketed through the majors, it is impossible to know whether this shift in control over production also created a shift in control over sales, thus stimulating price competition. In addition, the collapse of the deep-gas market since 1982 has forced many independents to sell acreage to the majors to avoid bankruptcy and has driven others out of business entirely.

Regardless of any progress made by the independents, however, other factors continue to inhibit competitive pricing. The pipelines still buy gas from their own affiliates, for instance, a practice that tends to push prices upward. And most of the major producing states, including Texas, Oklahoma, Louisiana, Kansas, and New Mexico, ration production from gas wells in the name of conservation, much the way OPEC restricts oil production to limit supplies and prop up prices. (Industry critics find it ironic that the major producers fiercely condemn government regulation when it holds prices down but remain silent when it props them up.)

If today's market were truly competitive, two years of oversupply would surely have driven average prices down. For the independents, it has. Lacking the financial muscle to survive long court battles, they have watched helplessly as pipelines have unilaterally cut the prices guaranteed by their contracts. But the major oil companies are fighting such moves in court, keeping the price of new gas from plummeting, as it logically should.

A similar dynamic prevails in the market for new supplies. As of October, the NGPA ceiling for new gas was $3.52. Some of the independents, with creditors nipping at their heels, were selling for $2.50 to $3.00, to the handful of pipelines that needed gas. But, according to Ted Eck, chief economist at Amoco, none of the majors was willing to sell significant volumes at this price level. A few long-term contracts at about $3.25, with automatic escalators up to $4.00 by 1985, were being signed—but even at $3.25 few majors were involved. "There are published surveys showing the pipelines simply have to buy more gas before the end of 1985, to fulfill their contracts," Eck explains. "So the market knows the pipelines have got to be much more aggressive buyers in the future. Sellers aren't willing to sell for $2.50 now if they think they can get $4.00 [then]." Since the majors have financial cushions to fall back on, only the independents have been forced to accept the dictates of a glutted market.

In a competitive market, gas prices would surely fall for *all* producers. Gas would replace most or all foreign oil in industrial and utility boilers with dual capacity, and it would compete mainly with domestic oil in the residential and commercial markets. Since most gas is cheaper to produce than oil, more and more of it would be produced, replacing oil, until the cheapest gas had been exhausted. At that point, as more-expensive gas was produced, gas and oil would reach price equilibrium, with consumers paying approximately the same amount per delivered unit of energy for each fuel. The amount of each resource that was consumed would be governed by the amount producible at that price. In the process, the price of gas would be the determining factor, dragging the price of oil down to its level.

In a noncompetitive market, however, just the opposite would happen. Gas prices would rise to the price of foreign oil and stay there. The price of gas would bear no relationship to the cost of production, as it would if producers were genuinely competing by offering their lowest price. Instead, it would be set by the cost of a competing *fuel*— which would in turn be set by a foreign cartel. Producers would be richer, consumers poorer, and the nation as a whole would continue to send $60 billion or more overseas every year for foreign oil.

Bob Hefner explained why oil companies would prefer to see oil determine the price of gas, rather than the other way around, in testimony before the House Subcommittee on Fossil and Synthetic Fuels last April. The key, he argued, is their desire to keep oil consumption high—which means keeping gas consumption low.

In a market-shares battle between oil and methane, integrated energy companies will always make more money selling oil than natural gas. Multinational oil producers make money not only when they discover and sell crude oil but also when they ship it, when they refine it, and when they market it at the wholesale and retail level. In contrast, these companies typically have no economic interest in the pipelines that transport nor in the distribution companies that market their natural-gas supplies. The revenues to be earned from selling gas cannot begin to compare with the revenues to be earned from selling oil.

Let me give you two concrete examples. Exxon's revenues in 1981 for sale of crude oil were over $14 billion. Revenues for refining and marketing of petroleum products totaled nearly $86 billion. Natural-gas revenues were only $2.5 billion. Exxon owns almost as many BTUs of natural gas as of crude oil. Yet it would take a forty-fold increase in the price of natural gas for Exxon's revenues from the sale of natural gas to equal revenues from production, refining, and marketing of oil. . . . Shell's revenues are twelve times as much from the production, refining, and marketing of oil as from the sale of natural gas. Shell, like Exxon, is primarily an oil, not a natural-gas, producer, and their corporate strategy—just as mine would be if I were they—will inevitably be to maximize oil markets and oil production, not natural gas.

Congressional investigations of the behavior of major oil companies during the oil shortage of 1973-1974, detailed by John M. Blair in *The Control of Oil*, revealed how far the majors will go to control prices and limit competition. When they lost control over foreign production to OPEC, refining and sales became the majors' high-profit sectors. To this end, they had to squeeze independent refiners and gas stations out of business, so as to limit price competition. They were remarkably successful. It is hard to imagine why they would not employ similar tactics in the gas business.

The future of natural-gas pricing is now in the hands of Congress. As matters stand, most new gas—some 50 to 60 percent of our total gas supply—will be decontrolled on January 1, 1985. A majority in Congress appears to support repeal of the Fuel Use Act; some action to curb the impact of automatic price escalators, which threaten to force gas prices even higher in 1985; some moderation of take-or-pay contracts; and stricter regulation of pipelines' purchasing habits. The Reagan proposal to decontrol old gas, however, has revived one of the great ideological struggles of the past forty years. During the last battle, in 1978, decontrol of new gas passed the House by one vote, after seventy-five weeks of bitter debate.

In the past, the oil and gas industries have always united behind decontrol. This time they are split down the middle. Old-gas decontrol would create a windfall estimated at $40 to $100 billion for the majors, who control more than 70 percent of all old gas. But as old gas rose in price, new gas would fall. Thus, most of the independents, who produce more than 70 percent of all new gas, oppose the bill. Most utilities and pipelines, fearing higher prices, also oppose decontrol.

Leading the fight against decontrol of old gas is the Citizen/ Labor Energy Coalition, which claims to represent some 300 organizations. CLEC has drafted its own measure, cosponsored by 154 representatives and seventeen senators. It would extend controls on new gas until 1987 and mandate rates lower than those under the current schedule. The bill would also allow take-or-pay purchases to be reduced to 50 percent of gas produced for three years; control deep gas at 150 percent of new-gas prices; and force tougher regulation of pipelines. Last October, the chairmen of the House Energy and Commerce Committee and its subcommittee on fossil and synthetic fuels, John Dingell, of Michigan, and Philip Sharp, of Indiana, joined forces to sponsor a compromise version of the CLEC bill that preserves most of its important features.

In Congress, the battle is not Republicans versus Democrats but producing versus consuming states. Since President Reagan will veto any recontrol of new-gas prices, as in the CLEC bill, that issue is moot. The real issue is decontrol of old gas: should Congress allow market prices, somewhere between $3.00 and $3.50, for gas that was profitable ten years ago at 46 cents?

The argument for decontrol of old gas turns on the assumption that higher prices will bring forth new supplies—gas that was either deliberately held off the market in anticipation of decontrol, or gas that would not be profitable at regulatory-ceiling prices, which now average $1.45. But since decontrol would bring the price of old gas up and that of new gas down—and thus discourage new drilling—this does not seem to be a wise bargain. Drilling is already at a virtual standstill, thanks to the gas glut. As a result, much of the industry expects another temporary shortage in 1985 or 1986. Incentives to produce more old gas would help, but incentives to produce new gas would help more, particularly in the long run.

Fortunately, there is a simple way to achieve both incentives. Old-gas producers might be allowed to charge higher prices if they could prove a well would not continue producing at profitable levels without expensive enhancement measures (such as pumping water down the well to raise its pressure). Federal regulations already permit this, if the buyer agrees. The mechanism might need some fine tuning, but such adjustments would be infinitely preferable to deregulating all old gas.

The key issue in the debate over decontrol of old gas, however, will be price: will the decontrol of old gas drive average prices up or down? The Energy Information Administration, created by Congress to provide independent analysis of the energy industry, says up. The oil companies and the DOE say down. They use the analogy of the trucking and airline industries, where deregulation unleashed fierce competition and brought consumers genuine bargains. Unfortunately, however, given the structure of the gas industry, a more appropriate analogy is oil-price decontrol. When President Carter initiated oil deregulation in 1979, prices more than doubled, to OPEC levels, costing the average American household $700 over two years. Most economists expect gas to follow a similar pattern, with prices rising to the level of residual oil. Utility and pipeline executives, who have every reason to understand gas pricing, agree—and give this as their reason for opposing decontrol. As they are quick to explain, any increase in gas prices will cut demand and guarantee that we continue to import great quantities of oil. In addition, rising prices for old gas will hand a windfall to the major oil companies and seriously injure the independents, futher undermining competition as power shifts back toward its traditional locus.

The debate in Congress will likely turn not on the broad, long-range interests of the nation but on the narrow, immediate interests of congressional constituencies—both voters and campaign contributors. During the 1982 campaigns, oil and gas political-action committees gave more than $7 million to congressional candidates, and they will no doubt be similarly openhanded this time around. On the other side, the Citizen/Labor Energy Coalition has unleashed what may be the largest genuine grassroots effort since the anti-ERA campaign; its activists claim that they knock on 50,000 doors a day. The CLEC campaign has had such an obvious effect on congressmen from consuming states that the Natural Gas Supply Association, which represents the major oil companies, has spent $3 million on its own door-to-door effort in support of decontrol.

This squeeze between voters and campaign contributors has engendered some perverse behavior, as was visible in the Senate Energy Committee last summer, while it rewrote the basic Reagan proposal. Just to get the bill out of committee with no recommendation, Republican Chairman James McClure, of Idaho, had to exempt several states from its key provisions. Frank Murkowski, a Republican from Alaska, supports decontrol in every state but his own, where most gas is cheap, purchased under contracts signed long ago, and where prices would soar if gas were decontrolled; before Murkowski would vote to release the bill, the committee had to exempt Alaska. Bennett Johnston, a Democrat from Louisiana, likewise approved the bill, but only after the committee had eliminated a provision decontrolling direct sales by producers to buyers (bypassing the pipelines); most such sales occur, to no one's surprise, in Louisiana. Malcolm Wallop, a Republican from Wyoming, also gave Chairman McClure his vote, but only after small producers who want to sell their gas collectively were granted immunity from the antitrust laws; Wyoming is, of course, home to many independents. Even Spark Matsunaga, a Democrat from Hawaii, which neither produces nor consumes much gas, demanded a price for his vote: DOE funding for two pet energy projects in his state.

If a majority in Congress were to step back and analyze the issues objectively, oblivious of political pressures, the picture would be quite clear. Given the large supply of gas now known to exist, national policy should strive to meet four simple objectives. The government should ensure that high volumes of gas are produced. It should eliminate all restrictions on the use of that gas, such as the Fuel Use Act. It should convince consumers large and small that gas is plentiful and will continue to be so. And it should prevent anticompetitive features of the energy industry from holding prices at artificially high levels.

How can these objectives be achieved? A first step would be to repeal the remaining clauses of the Fuel Use Act. Similarly, to bring prices down, partial relief should be granted buyers saddled with take-or-pay contracts and automatic price escalators. Pipelines should be held to stricter standards in their purchasing habits, and old gas should remain under price controls.

Even if these steps are taken, however, they will lead us only partway toward a solution. The key to stimulating use of our abundant gas supplies is keeping gas prices somewhere near the costs of exploration and production—and thus below the price of OPEC oil. In an anticompetitive industry, measures to this effect can take any of three directions. The first is to set prices at the cost of production plus a fair rate of return, as the FPC did from 1954 to 1978. The veto-bound CLEC bill would be a step in this direction, but only for three more years. In addition to its obvious political impracticality, however, this approach presents economic problems. When the government controls prices, it leaves production decisions in the hands of private corporations. As history has shown, this encourages producers to hold even profitable gas off the market, while they work to end price controls or invest their money in a resource whose prices are not controlled, such as oil. If the goal is a rapid increase of the supply of gas, price controls may be counterproductive.

The second strategy follows a precedent set even earlier: the use of federal antitrust powers to force genuine price competition among producers. Clearly this could bring about some progress: pipelines could be severed from their producing affiliates; joint ventures between large producers could be outlawed; oil companies and pipelines could be prosecuted for holding cheap old gas off the market while producing more-expensive new gas. But history has shown that successful antitrust prosecution of corporations as powerful as the oil majors is next to impossible. What the Justice Department wins, Congress takes away—or vice versa.

The third strategy is one that has been pursued successfully by other developed nations, including Great Britain and Norway: the establishment of a government energy corporation to compete in all sectors of the industry. The public corporation becomes a yardstick against which to measure the behavior of private firms. When the government believes prices are being inflated without reason, it simply sells its own supplies for less, undercutting the competition. In this way the Bristish government forced major oil companies to drop their prices for North Sea gas by almost 50 percent. Such a solution may appear even less likely than the first two in today's political climate. But if we accept the limitations of today's political climate, no solution is possible.

If the price issue were settled, the government could then move on to the question of consumption. Even after those industries and utilities with dual fuel capacity had switched from oil to gas, we would still be importing perhaps 3 to 4 million barrels of oil a day. To spur additional conversions, the government must lead.

Perhaps the most impressive inroads on the use of foreign oil could be made by converting automobiles to compressed natural gas. Experience in the United States and abroad proves that CNG works, particularly for business and taxi fleets, in which drivers do not normally travel more than 150 miles at a stretch. CNG tanks are too bulky for long-distance driving, although many cars are equipped to burn either CNG or gasoline, at the flip of a switch.

Because oil has to be refined to create gasoline, natural gas should always be the cheaper of the two automotive fuels. In British Columbia today, it costs half what gasoline does per mile driven. Those who are not familiar with compressed natural gas often assume that it is dangerous and can lead to catastrophic explosions. But in reality, it is safer than gasoline. CNG tanks are no more susceptible to damage or explosions than scuba tanks, which are filled to the same pressure. The Ministry of Universities, Science, and Communications of the Province of British Columbia, which is directing an ambitious conversion program, explains why CNG is safe:

CNG has four big safety advantages over gasoline or propane. First, it's lighter than air. If it leaks, it just rises and dissipates (gasoline and propane will puddle and build up a hazard). Second, it takes a temperature of 700°C or better to ignite. Gasoline ignites at a much lower temperature (450°C), and so does propane. Third, it takes specific CNG-to-air percentages—four to fourteen—for combustion. That is a narrower ratio range than gasoline's. Fourth, CNG tanks are built, by law, to hold extreme pressure—and are tested at 34.5 MPa (5,000 pounds per square inch) or higher depending upon the tank. That is more than 1.6 times the pressure to which they are filled. The tanks are heavy, rugged, and well-secured. They are less susceptible to collision damage than ordinary gasoline tanks. The American Gas Association found no natural-gas fires or explosions in 180 rear-end collisions with CNG-equipped vehicles.

According to a publication of the federal government in Canada, not one injury or death has been caused by compressed natural gas in thirty years of use in Italy, where some 250,000 vehicles burn the fuel. New Zealand, as well as Canada, is shifting a significant number of its vehicles from gasoline to methane.

Consumers will not pay $1,800 to convert a car to CNG, however—nor will they buy new autos specially designed to burn natural gas—until the fuel is widely available. Installation of natural-gas compressors and dispensers (both at gas stations and at home, in one's garage) could be encouraged through tax or other incentives, as they are in Canada. Consumers must also be convinced that plenty of natural gas will be available for many years. The government can play a crucial role here, not only in reassuring the public but also in helping industry develop the technology needed to exploit our huge geopressured gas reserves. Washington spends billions every year on nuclear power and synthetic fuels, both of which now appear to be dead ends, at least for the foreseeable future. Surely geopressured gas is worth a few million R&D dollars a year.

Besides going a long way toward resolving our energy problems, natural gas should be considered as a response to our environmental problems. Other than the sun and the wind and the sea, gas is our cleanest source of energy. In the long run, the wisest solution to our acid-rain problem may be full or partial substitution of natural gas for high-sulfur coal in midwestern boilers. Even partially completed nuclear plants could be converted to gas at great savings, as advocated by Dayton Power and Light for the troubled Zimmer nuclear plant in Ohio. And the more gas we substitute for oil and coal, the less we contribute to the buildup of carbon dioxide in the atmosphere—the greenhouse effect, which is expected to raise temperatures on Earth within a matter of decades. Methane gives off much less carbon dioxide when burned than coal or oil.

The agenda outlined above is obviously an ambitious one. Even if only a small part of it is fulfilled, however, natural gas could help solve some of our nation's most intractable problems. Consider the following scenario:

As take-or-pay contracts are interrupted, gas prices decline and consumption rises. Spurred on by renewed demand, wildcatters start drilling again, and supplies increase. As more and more of the nation's depths are explored, potential-reserves figures begin to rise. With significantly more gas moving through pipelines, the transmission cost per thousand cubic feet drops, keeping consumer prices steady even if wellhead prices creep upward. Acid rain begins to abate, as formerly under-used pipelines to the Midwest fill up with gas. Demand for foreign oil falls, reducing the dollar drain, improving the nation's trade balance, and keeping OPEC prices soft. In decades to come, as drillers go deeper and deeper for gas and prices rise, the nation shifts to renewable energy sources like solar power. Looking back, natural gas is seen to have served as a bridge to the new technologies—a function once envisioned for conventional-nuclear-power plants.

To many, this scenario may seem too rosy to be credible. But natural gas is already well on its way to becoming the world's primary source of energy. In 1979, two European scientists, Dr. Cesare Marchetti, and Dr. Nebojsa Nakicenovic, of the International Institute for Applied Systems Analysis in Austria, did a lengthy study of world energy consumption; they suggested that gas will soon displace oil as the number-one fuel.

In the Soviet Union, gas production grew by 50 percent from 1975 to 1980. In the Middle East, Saudi Arabia and some of its neighbors ship their oil overseas while building their own economies on gas. The inevitability of gas is perhaps best illustrated by Japan, which has almost none of its own. Because of Japan's severe air-pollution problems, most of that country's plans for expanded generation of electricity in recent years have depended upon liquefied natural gas—all of it imported.

Here in the United States, of course, we have the capacity to resist the inevitable. Ten years ago, President Nixon announced Project Independence, but we continue to import millions of barrels of oil every day. By 1975, many energy experts understood that nuclear energy was a failed experiment, but we continue to build nuclear-power plants. Ten years from now, will we still be building nuclear plants and importing foreign oil, while the solution to our problems languishes in the ground?

AN OLD FAVORITE RETURNS[2]

Although little noticed at the time, the visit had special significance. Last Dec. 12, Martha Hesse, chairman of the powerful Federal Energy Regulatory Commission—which oversees natural gas and electrical utilities in the United States—flew to Ottawa. The occasion: a meeting with Roland Priddle, chairman of Canada's National Energy Board, which controls energy exports and imports. Priddle declined to say what they talked about during the unusual meeting of the two energy czars. But insiders say that the visit underscored the growing importance of Canadian electricity and natural gas for the United States. In fact, Canada's troubled natural gas companies are already reaping the benefits of America's growing hunger for imported energy. Although prices remained stagnant, gas exports to the United States grew by 30 per cent in 1988, hitting 1.3 trillion cubic feet, the highest level in history. And the National Energy Board is predicting another nine-per-cent increase this year.

The long-term picture is looking even rosier. Experts say that gas use is poised to make a comeback in the United States, where it is viewed as a cheaper, cleaner and more dependable alternative to foreign oil. Western Canada is afloat in natural gas, which many companies have left stored in reservoirs under the ground rather than selling at current depressed prices. But existing pipelines can barely accommodate the 40 per cent of total Canadian production that is exported to the United States. However, by 1991, several huge cross-border pipeline projects that are now seeking regulatory approval will expand the export capacity.

American gas producers are fighting to keep the Canadian product out of their home market. But insiders say that president-elect George Bush has no intention of locking out imports—especially since the signing of the Canada-U.S. Free Trade Agreement, which forbids minimum export prices or taxes and prohibits either country from restricting exports except during emergencies. There is also political pressure by the heavily populated northeastern states for Washington to open the floodgates

[2]Reprint of an article by John DeMont, *Maclean's* staffwriter. Reprinted by permission from *Maclean's*, V. 102:26. Ja. 9, '89. Copyright © 1989 by *Maclean's*.

to Canadian fuel. Said William Martin, a former U.S. deputy secretary of energy with close ties to Bush: "The Free Trade Agreement will help give American end-users confidence that there is a secure supply of natural gas to fulfil their needs at competitive prices."

Indeed, only weeks after Prime Minister Brian Mulroney's re-election with a mandate to proceed with free trade, a chief judge of the U.S. Federal Energy Regulatory Commission endorsed separate pipeline plans by consortia led by TransCanada PipeLines Ltd. of Toronto and Noverco Inc. of Montreal. Combined, the two projects will be able to pump more than 900 million cubic feet of Canadian gas each day into the energy-hungry northeastern states—increasing U.S. gas imports from Canada by an estimated 25 per cent.

By hammering out a settlement among the 37 competing parties, the U.S. regulators managed to sidestep a complex approval process that could have dragged on until 1993. Instead, the Canadian-led consortia will file their final proposals to the U.S. regulatory commission on Jan. 17, and production on the new lines could begin as early as November, 1990.

As well, other plans are afoot to ease the pipeline bottleneck. Five consortia have begun the long process of seeking regulatory approval to build a new pipeline that would move more Canadian gas into California—the fastest-growing gas market in the United States—where the two major existing pipelines are regularly full to capacity. And Shell Canada Ltd. and Esso Resources Canada Ltd. recently applied to build a new multibillion-dollar pipeline needed to connect arctic gas to pipelines in southern Canada.

Meanwhile, on Nov. 30, TransCanada PipeLines received National Energy Board approval for a $567-million expansion, which will allow it to sell an additional 114 billion cubic feet of natural gas during 1989 to new markets in Canada and the United States. And the Toronto-based gas utility is about to file a new application with the National Energy Board to build further facilities along its existing system to ship an additional 201 billion cubic feet of gas a year to the United States. Said James Cameron, president of the pipeline division of TransCanada PipeLines: "We haven't had growth of this nature for years."

Even so, beleaguered western Canadian gas producers say that they hope the export boom will continue long enough for gas prices to start increasing again. Since 1984, depressed prices and a North American surplus of the fuel have forced Canadian producers to leave large amounts of gas stored in the ground. Now, such companies as Czar Resources Ltd. of Calgary are scrambling to find all the oil they can to export to the United States, where prices are higher than in Canada—even though prices are still less than half of what they were five years ago. Said Czar president Robert Lamond: "Most of us are just trying to expand our volumes and stay in business until the prices eventually come around."

Canadian gas companies can likely count on support from Bush, who described natural gas during the recent election campaign as "America's energy ace in the hole" and who is under pressure to reduce acid rain and other forms of air pollution. According to Bush's adviser Martin, natural gas holds great promise for reducing the United States' rising dependence on foreign oil. Martin says that the Organization of Petroleum Exporting Countries' recent agreement to limit production and drive up the price of crude next year may make cheaper-priced gas look even more enticing. As well, public pressure could force coal- and oil-burning utilities to switch to natural gas as a way of reducing air pollution. Said Martin: "Gas is something which Bush thinks America should take a good look at." Canada's gas company officials are expressing optimism that the new U.S. administration will look to the North.

SCRUBBER SCRAPPER[3]

Good news for humans and other species worried about acid rain, beyond the shadow of a trout: this year several utilities are breaking ground by adapting an old technology to burn coal cleanly. The process, called atmospheric fluidized-bed combustion, offers the prospect of cleaner emissions at lower cost than can be achieved with the methods now used. It could become the mainstay of new and refurbished coal-fired plants.

[3]Reprint of an article by Edward C. Baig, *Fortune* staffwriter. Reprinted by permission from *Fortune*, V. 113:63–4. Ap. 14, '86. Copyright © 1986 Time Inc. All rights reserved.

Since World War II most coal has been burned in pulverized form in hot-air furnaces at about 2,500° F., producing not only steam that drives turbines but also nitrogen and sulfur oxides, both linked to acid rain. Utilities have been able to capture up to 90% of the sulfur emitted by installing cumbersome devices called scrubbers that clean the exhaust gas before it goes out the chimney. But scrubbers are expensive—they can amount to one-third the cost of building a power plant—and virtually ineffective against nitrogen oxides. Most scrubbers also produce a wet sludge that is difficult to dispose of.

In conventional fluidized-bed boilers, different-size chunks and grades of coal are mixed with inexpensive limestone. Air is forced in, sometimes at such high velocity that the coal and limestone float around like ping-pong balls. The limestone reacts chemically with the coal to capture as much sulfur as a scrubber would. The result is a dry waste product, which may have some use as a building material. Even better, the temperature remains about 1,000 degrees lower than in a pulverized furnace, reducing the quantity of nitrogen oxide produced.

The fluidized-bed concept was born in Germany in the 1920s, but the utility industry did not try to make the process economical at a power plant until the 1970s, when oil prices hit the roof. A pilot plant in Paducah, Kentucky, run by the Tennessee Valley Authority since 1982, demonstrated the commercial potential of the technology for coal.

Now the idea is being applied to larger plants. This summer, at its Black Dog facility in Burnsville, Minnesota, Northern States Power (1985 assets: $4.1 billion) will "light off" a $52-million fluidized-bed boiler. The Colorado Ute Electric Association ($1.2 billion) is spending $87 million to install a fluidized-bed boiler and triple the capacity of its Nucla plant by August 1987. The TVA ($21.5 billion) has broken ground on a brand-new $205-million power plant, also in Paducah. And Wisconsin Electric ($2.3 billion) plans to retrofit four separate units in Oak Creek, Wisconsin, with the new technology. When the $380-million project is finished in five to six years, Oak Creek will be the largest fluidized-bed combustion station in the world.

Says Charles Linderman, manager of the fossil fuels program at the Edison Electric Institute, an industry trade association in Washington, "These plants can lead us to the environmentally benign utilization of coal." Nearly all of the utilities doing the pioneering have been helped along by money from the Electric Power Research Institute, an industry-funded outfit in Palo Alto, California; from other private companies; or from the federal government. Industry experts say no further subsidy for building conventional fluidized-bed plants is needed.

A government and industry anti-acid-rain program that President Reagan discussed in March with Canadian Prime Minister Brian Mulroney may include funds to help develop so-called pressurized fluidized-bed systems, in which the air is forced upward under even greater pressure. This is a more efficient way to burn coal, but as yet the utilities that are working on the systems— Wisconsin Electric and Ohio Power Co.—have not proved they can harness the gases produced.

COAL-FIRED POWER PLANTS FOR THE FUTURE[4]

The invention of the incandescent light bulb by Thomas A. Edison in 1879 created a demand for a cheap, readily available fuel with which to generate large amounts of electric power. Coal seemed to fit the bill, and it fueled the earliest power stations (which were set up at the end of the 19th century by Edison himself). As more power plants were constructed throughout the country, the reliance on coal increased. Since World War I coal-fired power plants have accounted for about half of the electricity produced in the U.S. each year. In 1986 such plants had a combined generating capacity of 289,000 megawatts and consumed 83 percent of the nearly 900 million tons of coal mined in the country that year. Given the uncertainty in the future growth of nuclear power and in the supply of oil and natural gas, coal-fired power plants could well provide up to 70 percent of the electric power in the U.S. by the end of the century.

Yet in spite of the fact that coal has long been a source of electricity and may remain one for many years (coal represents about 80 percent of U.S. fossil-fuel reserves), it has actually never been

[4]Reprint of an article by Richard E. Balzhiser and Kurt E. Yeager, writers for Scientific American. Reprinted by permission from Scientific American, V. 257:100–7. S. '87. Copyright © 1987 by Scientific American. All rights reserved.

the most desirable fossil fuel for power plants. Coal contains less energy per unit of weight than natural gas or oil; it is difficult to transport, and it is associated with a host of environmental issues, among them acid rain. Since the late 1960's problems of emission control and waste disposal have sharply reduced the appeal of coal-fired power plants. The cost of ameliorating these environmental problems, along with the rising cost of buliding a facility as large and complex as a coal-fired power plant, have also made such plants less attractive from a purely economic perspective.

Changes in the technological base of coal-fired power plants could restore their attractiveness, however. Whereas some of these changes are evolutionary and are intended mainly to increase the productivity of existing plants, completely new technologies for burning coal cleanly are also being developed. These technologies seek to make cost-effective environmental control, flexibility in the type of coal burned and reduced construction time inherent features of future coal-fired power plants.

To appreciate the significance of such advances in coal-combustion technology, a brief review of the operation of a conventional coal-fired power plant is in order. A modern power plant burns coal in a boiler: a huge box-shaped vessel whose inner walls contain tubes in which water is converted into steam. The coal is finely ground before it is injected into the boiler so that it burns with the high efficiency of a combustible gas. A typical large boiler consumes 500 tons of pulverized coal every hour in order to produce 6.4 million pounds of steam—enough to generate a million kilowatt-hours of electricity. At the same time the boiler gives off about 3.5 million cubic feet of flue gas, or exhaust.

The steam passes through a superheater, where its temperature and pressure are increased, before it drives a high-pressure turbine. The rotation of the turbine shaft provides the mechanical energy that is converted into electricity by a generator. To increase the overall energy-conversion efficiency, the steam leaving the turbine is usually reheated and sent back to drive one or two lower-pressure steam turbines before it is cooled, condensed and pumped (as water) to the boiler to begin the cycle once again.

In addition to coal-feed mechanisms, boilers, turbines, condensers, pumps and generators, power plants also depend on elaborate cooling, emission-control and waste-handling systems. All these main and adjunct systems must be designed to operate at high reliability for a lifetime of 40 or more years under load conditions that may vary daily between 20 percent of the plant's design capacity and its maximum power output. The capital costs of the equipment in a typical 1,000-megawatt plant can easily exceed $1 billion.

The efficiency with which the heat released from the burning of coal is converted into electricity, which was as low as 5 percent before 1900, had reached about 40 percent by 1967. In other words, in the space of about 70 years an eightfold reduction in coal consumption per unit of generated electricity was achieved. This progress was reflected in the cost of a kilowatt of new generating capacity: in 1920 it was about $350 (in 1967 dollars) and in 1967 it had dropped to $130. The cost of residential electricity service also dropped from more than 25 cents per kilowatt-hour to two cents per kilowatt-hour over the same period.

Beginning in the 1960s, however, the pace of improvement began to slow. This trend suggested that conventional power-plant designs had approached the inherent limits set by the laws of thermodynamics and the properties of the materials from which boilers and steam turbines were made. By the 1970's the diminishing technical returns were also joined by new economic and institutional forces, namely rapidly escalating capital costs, slow growth of demand, stricter emission-control requirements and extended licensing and construction schedules. The result has been an abrupt end to the historic trend of declining real cost for electricity from coal. Indeed, a kilowatt of new generating capacity costs more today (in constant dollars) than it did in 1920.

Measures to control gaseous, liquid and solid wastes have had the greatest impact on the cost of coal-fired power plants in the past 20 years. A new plant's pollution-control systems now account for as much as 40 percent of the capital outlay and 35 percent of operational costs. The most technically and economically visible element of emission control in today's pulverized-coal plants is the flue-gas-desulfurization system, often called a wet scrubber. The scrubber removes the oxides of sulfur that are a major gaseous pollutant resulting from the burning of coal.

Wet scrubbing is a simple concept but is difficult and expensive to carry out in practice. An alkaline substance, usually lime or limestone, is mixed with water to form a slurry that is sprayed into the flue gas. The sulfur oxides in the flue gas are absorbed by the slurry and precipitate out of the liquid as inert calcium sulfite or calcium sulfate (gypsum). Gypsum, if it is pure enough, can be easily disposed of or sold as building material. Alternative but more expensive scrubbing systems are able to transform the precipitates into sulfuric acid or elemental sulfur, which can then be sold on the chemical market at a higher price. Scrubbers have been required on all pulverized-coal power plants whose construction began after 1978. As a result the U.S. electric utility industry is operating more wet-scrubber systems today than the rest of the world combined.

In a new plant the scrubber typically costs between $150 and $200 for each kilowatt of generating capacity. Because of the difficulty in fitting scrubbers into old power plants not originally designed to accommodate them, a scrubber for an existing plant costs between 10 and 40 percent more than a scrubber for a new plant—assuming that the installation goes smoothly. Regardless of the age of the plant, the cost of operating the scrubber is high. Scrubbers create huge amounts of sludge waste that must be put in holding ponds and landfills, constituting an environmental side effect in itself. A 1,000-megawatt power plant that burns coal with a sulfur content of 3 percent, for example, produces enough sludge in one year to cover about a square mile to a depth of one foot.

Wet-scrubbing systems also require large amounts of water (about 1,000 gallons per minute for a 1,000-megawatt plant), and they often suffer from plugging and fouling of equipment and corrosion of duct-work. These problems add to operating costs and reduce the overall reliability of the system. Finally, scrubbers extract a penalty of between 3 and 8 percent of the power plant's energy output simply to run pumps and fans and to reheat the flue gas in order to prevent corrosive condensation in the chimney.

The widespread implementation of scrubber technology in the U.S. has not been easy or cheap. The reliability of early units was considerably less than that of the power plant as a whole; hence they required components that were either redundant or had been engineered with large margins of tolerance. Some of the difficulties associated with the installation and operation of scrubbers can be attributed to the fact that the technology was prematurely brought up to commercial scale. Only now, after a quarter-century of experience, have scrubbers approached an acceptable level of reliability.

Coal-fired power plants have become more expensive to build not only because of emission-control systems but also because construction costs themselves have skyrocketed. Even after adjusting for inflation, new coal-fired generating capacity is three times as expensive today as it was in 1970. Over the past 15 years the economies of scale that are to be gained from constructing large plants have been offset by large cost increases. In part the increases reflect the high cost of financing prolonged construction delays.

The experience of Japanese utility companies demonstrates how big a difference construction delays can make. Japanese utilities typically deal more expeditiously than their American counterparts with the various regulatory and financial issues that often delay large construction projects. Whereas the Japanese are able to bring power plants on line in from 35 to 40 months, American companies generally require from 50 to 60 months for plants of similar design. Consequently the cost of a single new large generating unit is comparable to the total assets of many U.S. electric utilities.

Utilities are therefore seeking to reduce the costs by turning to smaller, modular generating units, which can be transported to the sites of existing power plants and rapidly installed there to keep pace with the growth of demand. Because such units can be brought on line quickly, their capital cost can be recovered faster—even if the rate base remains constant. Installing new units only when new capacity is needed can result in a net saving of as much as $200 per kilowatt, in spite of the fact that smaller units sacrifice economies of scale.

As an alternative to building new generating capacity, utilities have been refurbishing old plants to improve their performance and extend their lifetime. This strategy is certainly less costly than building a new plant. The trend is reinforced by the fact that plants now completing three decades of service are not obsolete. In some cases they may even be more efficient than new plants, since they are not burdened by scrubbers. Indeed, older plants are becoming a larger fraction of our total electricity-generating capacity. In 1970 only 2 percent of U.S. generating capacity was more than three decades old. By the end of the century 30 years will be the average age of a coal-fired power plant.

Utilities are also looking at ways to lower operating costs. Early warning of the deterioration of critical system components is essential to prevent the loss of generating capacity. Hence the continuous monitoring for incipient failure is becoming an important part of maintenance procedures. Such monitoring keeps track of the natural processes of wear, corrosion and erosion, making it possible for plant operators to institute corrective measures in time to avoid forced electricity outages. The value of such efforts can be appreciated by recognizing the fact that an idled 1,000-megawatt coal-fired plant may cost the utility as much as $1 million for every day it is out of service, largely because the electricity must be bought from other, more costly sources.

Escalating costs for the transportation and handling of coal as well as for ash disposal have also made the quality of coal (determined from its content of moisture, sulfur and other minerals) an important consideration in improving plant performance. Although low-quality coal may well be cheaper per ton than high-quality coal, more of it must be burned to generate the same amount of electricity. The cost of shipping a greater volume of low-quality coal may offset the cheaper price. Low-quality coal typically generates more waste than higher-quality coal, leading to higher costs for waste disposal. Finally, the composition of low-quality coal is more variable than that of the better grades, making it difficult to "tune" a plant's systems to achieve the greatest possible efficiency: the systems have to be adjusted to cope with the worst quality of coal expected in a load.

Existing plants can improve the coal or at least maintain a uniform quality by removing some of the impurities, such as sulfur-containing minerals, before the coal is burned. This is done at facilities that crush the "dirty" coal and then separate the coal from the mineral impurities by virtue of differences in specific gravity or other physical properties.

In spite of such measures for improving the performance of existing coal-fired plants, the U.S. may still need 150,000 megawatts or more of additional generating capacity by the end of the century, if demand for electricity grows at the projected annual rate of 2.3 percent. To maintain the competitiveness of coal in such an expanding energy market, utilities will have to adopt coal-burning technologies that are more cost-effective than conventional ones in three key areas: pollution control, construction and performance.

There are a dozen or more advanced technologies under development for burning coal cleanly and efficiently. Two of the more promising are fluidized-bed combustion (FBC) and coal gasification. In a typical FBC coal boiler, crushed coal is "fluidized" along with bits of limestone by supporting the particles on a strong rising current of air. The fluidized particles act as if they were in a boiling liquid, that is, they mix turbulently, ensuring a very efficient combustion process. Because boiler tubes are in direct contact with the fluidized particles, an FBC boiler can absorb much of the heat by conduction, which is more efficient than the radiant and convective heat transfer a conventional boiler relies on.

The greater surface area for heat transfer offered by the tubes in the fluidized bed also enables an FBC boiler to operate at lower temperatures than a conventional pulverized-coal boiler does, thereby minimizing the formation of oxides of nitrogen. (Whereas the temperature in a conventional boiler may reach 3,000 degrees Fahrenheit, in an FBC boiler it generally ranges between 1,450 and 1,600 degrees.) Moreover, the limestone mixed with the coal captures 90 percent or more of the sulfur released from the coal during combustion, because the lower operating temperature promotes the reaction between limestone and sulfur that produces calcium sulfite or sulfate. FBC tackles the pollutants produced by burning coal where they are formed: in the furnace.

FBC boiler design and operation are also less sensitive to variations in coal quality. The large amounts of melted ash that result when low-quality coal is burned in a pulverized-coal boiler often foul heat-exchange surfaces, reducing the boiler's efficiency and reliability. In an FBC boiler, however, coal is burned at temperatures below the melting point of ash, avoiding the problem. Because an FBC plant operator can afford to buy lower quality coal, he can in some instances substantially reduce operating costs.

FBC technology also lends itself to the design of small boilers that can be prefabricated as modular units. It is estimated that a plant built from compact, pressurized FBC boilers could require between 10 and 20 percent less investment capital than conventional plants having the same total generating capacity. The savings result from a reduced construction time and better capacity to expand the plant in response to an uncertain growth in the demand for electricity; planning is also made easier by the knowledge that these compact units can be installed quickly whenever the need arises.

An FBC boiler could also be added to an existing plant for a rapid increase in generating capacity. Indeed, the Northern States Power Company has converted an existing coal-fired boiler in a Minnesota plant into an FBC boiler. The objective is to increase generating capacity by 40 percent, increase fuel flexibility (the boiler can burn even municipal refuse), control emissions more thoroughly and extend the plant's operating life to at least 50 years.

Over the past 15 years the technology of power plants that rely solely on FBC boilers has advanced from small test and pilot-plant facilities to large demonstration plants. The Tennessee Valley Authority, the Duke Power Company and the Commonwealth of Kentucky are building such a plant with a generating capacity of 160 megawatts; the Colorado-Ute Electric Association, Inc., has begun operating a 110-megawatt FBC-based generating unit. If they are successful, these two projects along with the Northern States Power Company's project—a combined private-sector commitment of nearly $400 million—will reduce the economic risk entailed in the application of FBC boilers for the utility industry.

An alternative technology, which—in less sophisticated form—was actually commonplace in the mid-19th century, produces clean-burning gas from coal. The gas is suitable for lighting and heating purposes and in fact had an important role in the U.S. until it was displaced by natural gas just before World War II.

Utilities were initially interested in coal-gasification processes because they can produce a fuel for boilers that burns cleanly and thereby eliminates the need for scrubbing. It is now clear that coal gasification has a more significant advantage: turbines can be driven directly by the hot gases produced by the combustion of coal gas. This in turn makes it possible for the "waste" heat of the combustion products leaving the gas turbine to be captured and harnessed to produce steam for a steam turbine. Such an arrangement of coupled gas and steam turbines is called a combined cycle and is one of the most efficient means available to generate electric power.

Coal gas from which sulfur and particulate matter have been removed is an excellent gas-turbine fuel and, like natural gas, can be burned with very low emissions. The combined cycle's high efficiency compensates for the unavoidable efficiency losses that result from converting coal into a gas. Moreover, water consumption is lower than in a conventional coal plant because two-thirds of the power is produced by the gas-turbine system, which needs no cooling water to condense steam.

The viability of an integrated gasification/combined-cycle (IGCC) power plant has been demonstrated at the Cool Water site of the Southern California Edison Company. The Cool Water plant, which began operating in May, 1984, generates about 100 megawatts of electricity from a wide range of coal types; its emissions are comparable to those of a nearby natural-gas power plant. Sulfur oxides in its flue gas are kept well below Federal standards by means of an adjunct sulfur-removal system that extracts essentially all the sulfur originally contained in the coal to yield pure sulfur with a significant market value. Nitrogen oxide formation is suppressed by adding moisture to the gas before it is burned, so that the combustion temperature of the gas is lowered. Moreover, the coal residues left in the gasifier are melted, yielding an inert, vitreous material when cooled that meets California solid-waste standards.

In addition to its advantages in pollution control and efficiency, an IGCC plant can be constructed in phases, allowing generating capacity to be built up in increments. Such flexibility in construction minimizes the risk in capital investment associated with an uncertain growth of the electric load. The first increment of generating capacity, for example, might be provided by gas turbines, which, taking advantage of current low fuel prices, might be driven not by coal gas but by oil or natural gas. Then, as the demand for electricity grows, the addition of a waste-heat boiler and steam turbine would increase not only the capacity of the plant but also its efficiency. When oil and gas prices as well as load requirements make it economical to do so, a coal-gasification facility could then be constructed at the plant site.

The role of coal-fired power plants is a crucial question in the public-policy debate over the stewardship of natural resources, the protection of the environment, the provision of energy and the growth of the economy. Yet these considerations need not conflict: a look at the emerging technological options shows that emission controls and cost reductions can in fact complement one another. This principle was recognized in a joint U.S.-Canadian report on acid rain issued last year. Acting on the report's recommendations, Congress is currently considering the establishment of a major national initiative to demonstrate and deploy clean coal-burning technology. This initiative, which would augment private sector resources with Federal investments, is intended to ensure that the full range of coal-processing technologies— including FBC boiler and coal gasifiers—will be available for widespread commercial application in the 1990's. Even with such power-plant technologies in the offing, however, the growing demand for electricity could not be met without a collection of concurrent measures that emphasize conservation, end-use management and increased productivity from existing power plants.

Continuing economic and environmental considerations will eventually lead to technological developments extending well beyond those we have discussed here. Ultimately coal-fired power plants could evolve into fully integrated "resource refineries." Such facilities would adapt themselves to the local fossil-fuel and resource base to produce a mixture of electricity, heat, fuels and marketable products for the local economy. In addition to FBC boilers and coal-gasification facilities, an entirely new spectrum of technologies—linked by electronic diagnostic and control systems—would be incorporated at such plants to recover most of the by-products of coal processing.

There is no shortage of opportunities to improve both the economic and the environmental aspects of coal-based electric-power production. The timely exploitation of these opportunities, however, depends on whether the Government can enact broad and balanced energy and environmental policies that give the electric-utility industry the necessary incentives. Care must be taken that innovative coal-processing technologies are developed and implemented rationally, with the cooperation of utilities—in contrast to the way scrubber technology was introduced. This could be ensured if the associated costs and risks are minimized through well-planned design, testing and refinement of small-scale prototypes and demonstration plants before commercial-scale systems are finally introduced.

LOW-HEAD HYDROPOWER FOR LOCAL USE[5]

The repair of small dams and the rebuilding of hydroelectric generating equipment has recently attracted much attention. Although flowing water is a renewable resource and essentially a free fuel, many small hydroelectric units throughout the United States were abandoned over the last half-century as uneconomic. But increases in other fuel prices (especially oil, natural gas, and coal) have made it appropriate to take a new look at the potential for expanding the small-scale or "low-head" hydroelectric energy supply. At hydro facilities, the amount of energy that can be generated depends on two factors: the volume of water (flow) and the vertical distance the water falls (head).

Low-head hydropower represents a unique, valuable resource for many *local* communities. From the perspective of the United States or even from that of the individual states, legitimate institutional factors and invariable physical limitations make low-head hydropower a doubtful option for meeting future energy needs. Three different elements contribute to this: (1) technological changes in electric power generation have eroded the advantages of small-scale hydroelectric plants over the years; (2) the physical resource base available for hydropower is a limiting factor; and (3) competing uses for inland waters constrain the technical potential of this physical resource base. During the course of this discussion, Michigan will be used as an example because it has many of the constraints typical to modern expansion of low-head hydropower and because the situation in this state has already been analyzed in some detail. Additionally, Michigan ranks third in the nation for both the number of retired hydro dams and the amount of unused hydro-generating potential, and thus serves as a prime illustration of the pros and cons of hydropower. This discussion, however, is applicable to many other states and to the redevelopment of low-head hydropower in general. People living

[5]Reprint of an article by Peter Kakela, associate professor of Resource Development at Michigan State University; Gary Chilson, post-doctoral research associate at Michigan State University Department of Resource Development; and William Patric, graduate student at Michigan State University Department of Resource Development. Reprinted from *Environment*, V. 27:31–9. Ja./F. '85, with permission of the Helen Dwight Reid Educational Foundation. Published by Heldref Publications, 4000 Albemarle St., N.W., Washington, D.C. 20016. Copyright © 1985.

in states with plenty of water, but not much fall, can readily identify with Michigan. Those living in other states with greater topographical relief, or even more water, can expect low-head hydropower to make a large potential contribution to their state's supply of electricity. Residents of flatter or dryer regions than Michigan must obviously expect a smaller contribution. State regulations differ greatly, but all must comply with certain federal requirements.

Historic Trends

The electrical age began about a century ago. On New Year's Day in 1880, Thomas Edison unveiled his marketable incandescent light bulb. Within two years, the first commercial hydroelectric facility began to operate. A waterwheel on the Fox River in Appleton, Wisconsin, drove turbines on two generators with a total power output of 25 kilowatts. The energy was sold to two paper mills and one residential customer for lighting the new incandescent lamps.

Hydropower played a prominent role in total generating capacity during the first half-century of the electric era. As late as the 1930s, for example, it provided nearly 40 percent of the nation's total electricity requirements. Even in the late 1940s, hydropower was the primary electricity source in some states—for example, 65 percent in Maine, 68 percent in New Hampshire, and 83 percent in Vermont.

Because of its many virtues, hydropower was an important component of the rapidly expanding electrical industry; it was and still is clean, efficient, dependable, and it uses a free, renewable fuel. There are, however, some drawbacks. Dam sites are fixed in location. Many suitable sites for power generation are far removed from potential consumers (early technology for the transmission of power over long distances was very inefficient). The magnitude of potential power generation is also governed by the physical landscape, and the combination of hydrology and topography seldom yields as much electric generating potential as customers demand.

Early technologies for generating electricity from coal, oil, or gas also had problems. The most significant of these was that early thermal generation of electricity was not very efficient. Consequently, the first fossil fuel plants were small and had to be located close to customers. To make better use of the total system, these thermal units piped excess steam to nearby homes and businesses in so-called district heating (or city steam) systems. District heating improved the overall efficiencies of the early thermal units, thus helping them compete economically with hydropower and its free fuel.

Great technical advances have since been made in the efficiency of steam generation, reducing the amount of energy (Btu) required to produce a kilowatt-hour of electricity. For example, maximum efficiency rates for thermally generated electricity more than doubled between 1920 (17 percent) and 1965 (39 percent). Consequently, electricity in the mid-1960s could be generated with half as much coal as in 1920. This trend has now reached a plateau, and further improvements in thermal generating efficiencies are no longer expected.

The development of thermal electric generation was facilitated both by improved thermal efficiencies and by the relatively stable cost of coal. When inflation is discounted, the real price of coal even declined throughout the 1950s and 1960s, but rose in the 1970s. This was also true for oil and gas prices.

There is something of a chicken-and-egg question here: which came first, technological improvements or demand? Did improved technology and lower prices encourage the promotion of electricity, and therefore cause demand to grow, or was there a pent-up demand that stimulated technological developments? Both situations probably had an effect, but it is possible that improved technology may have allowed a little more electricity to be produced at lower prices than before and that more people were willing to buy the cheaper power.

At any rate, demand did grow, and the physical limitations of hydropower sites began to be of more and more concern. Any environment has a limited number of suitable hydro sites; there is not enough water in the right places to generate all the electric power currently demanded in this country. This is especially true in states where the majority of sites are low-head.

Economies of Scale

There is more to the issue than physical limits alone, however, because some of the previously operating hydroelectric generators have recently been decommissioned. To comprehend how the effects of size promoted the abandonment of small-scale hydro units, it is necessary to understand the relevant economies of scale.

As the demand for electricity increased, the number of companies selling power at first grew rapidly. From the 1880s—when the original Edison Electric Company was formed—to the early 1920s, more than 6,500 different companies became established and sold electricity to cities and towns across the country. In the 1920s, several financial and governmental policy changes included the decision to recognize electrical utilities as natural monopolies. As a result, the number of electric utility companies declined sharply, although sales continued to rise rapidly. Thus, beginning in the 1930s, fewer and fewer companies were supplying more and more power, and the size of electric companies grew enormously.

The bigger companies began to search for simple, comprehensive solutions to satisfy the ever-growing need for power generation. The operation of just one large plant is much easier to manage, schedule, control, and maintain (and therefore is less costly) than the operation of many small plants. Reduced labor costs provided the biggest savings, but construction costs are also smaller. As a result, the total costs for each unit (kilowatt-hour) of power generated decline as the size of the plant increases—a classic example of economies of scale.

Low-head hydropower has a serious disadvantage when it comes to economies of scale—low-head is small-scale. At hydro facilities, in order to increase the potential for generating electricity, either more flow or more head is needed. In terms of cost, however, an increase in head will produce cheaper power than will an increase in flow. A smaller volume of water falling from a greater height requires smaller turbines and, therefore, lower investment costs than a larger volume of water falling a short distance. Thus, production of the same amount of electricity with low heads and large flows (the situation in many of the eastern and central states) costs more per unit of power than if less water fell from mountainous heights.

The Shift to Thermal Power

The electric power industry chose to favor larger and larger thermal generating units for at least five reasons:

- the ability to meet the ever-growing demand for electricity;

- the economies of scale available;
- the convenience of "one comprehensive solution";
- the great improvements that had been developed in thermal generating efficiencies; and
- the periodic decrease of fuel costs.

This shift toward large thermal units led to the decline in the average real (deflated) price of a kilowatt-hour of electricity from the 1930s through 1972.

Consequently, electric utilities began to retire some of their smaller hydroelectric units in the middle of this century. Recent Federal Power Commission figures for the nation indicate that "over 770 hydro plants have been abandoned since 1940." The demise of small-scale hydropower in Michigan, for example, dramatically reflects this trend. Consumers Power Company retired or sold 44 hydro plants between 1950 and 1973. In 1963, Michigan's other large utility, Detroit Edison, decommissioned all five of its hydro sites. During the 1950s and 1960s, many other sites operated by investor-owned utilities, municipalities, and industries were closed.

A very different period began in the 1970s, however. A relatively complacent attitude toward energy supply was shattered by the 1973 oil embargo by the Organization of Petroleum Exporting Countries (OPEC). Since then, the tremendous increases in oil prices have caused the prices of other fuels to rise, as well as the costs of most products. As fuel prices soared during the 1970s, many people began to think once again about water power and its relative immunity to changes in the price of oil. In 1978, for example, Robert Loube, a Michigan economist, commented that

the price of oil for utilities more than tripled between 1972 and 1976. The price of coal more than doubled. The price of water, of course, did not increase. This shift in relative cost provides the incentives to re-evaluate Michigan's hydroelectric potential.

Physical Realities

Like Michigan, many states have plenty of water but simply lack mountains. This is a critical factor in determining the potential of hydropower. However, other physical factors also play a role. The volume of water flowing in a river may vary considerably over the course of a year, primarily because of seasonal fluctuations of rainfall and snowmelt. The holding capacity of most watersheds not yet developed for hydropower is also relatively small and in many cases may have been reduced even more by extensive urban development.

These physical factors are by far the most significant constraints on hydropower development. Many states, however, still have some dormant hydro potential, even though they may not be as well suited for hydropower production as the prime areas of New England and the Pacific Northwest.

It is important to recognize that even a limited hydro potential can provide a valuable, clean, renewable energy source, especially at the local level. For example, the renovation of a decommissioned dam could result in the supply of a significant portion of a small city's power requirements and net the city considerable cost savings. Michigan's most ambitious rehabilitation plan is a four-dam renovation project on the Boardman River near Lake Michigan. When completed, the dam will add almost 18 percent to the present coal- and oil-fired generating capacity of Traverse City.

Theoretically, some rural communities could become virtually energy self-sufficient by using hydropower, or could become net producers of electric power. Hydropower has its place in nearly every state, but only within the realistic context of its very limited availability.

Regulatory Constraints

Many regulations control the development of dormant waterpower potential. Hydropower comes under both state and federal regulations because most streams are considered interstate navigable waterways. There are at least 17 laws at the federal level and, in Michigan, there are 10 state statutes that might be invoked to directly or indirectly influence hydro developments.

When the specifics are considered, the hydro regulatory process is simpler than it first appears. The situation naturally varies from case to case, but there are only a few essential regulatory requirements for any hydro rehabilitation project. At the national level, the Federal Energy Regulatory Commission (FERC) gathers the numerous federal concerns under a single roof and handles the federal permit and license requirements. In the state of Michigan, only three legal steps are necessary before beginning a hydroelectric rehabilitation project: complying with the federal requirements, procuring a Michigan inland lakes and stream permit, and obtaining an agreement on fish passage.

The Inland Lakes and Streams Act represents the guts of Michigan's regulation of hydropower. The stated purpose of the Act is "to regulate inland lakes and streams, to protect riparian rights and the public trust . . . ." Most states have similar legislation and require a permit for any work performed below a lake's or stream's high-water mark. Consequently, a developer will probably be required to obtain a permit for even minimal work at an existing dam.

Sometimes the state regulatory agency may consider more than just the structural engineering aspects of a hydropower proposal and its impacts on water rights and the public trust. Based on a wide range of concerns—including those on environmental, recreational, aesthetic, and safety matters—the regulatory agency in some states might also request a full report on the environmental effects of the project as part of an environmental assessment process.

One of the most significant non-physical constraints affecting low-head hydro development in Michigan—certainly the most controversial and emotional constraint—is the issue of fish passage at hydroelectric dams. Regulating fish passage over dams is usually defensible for economic reasons. This is especially true in Michigan, where the fishery resource is considerably more valuable than the hydropower redevelopment potential. In 1980, for example, Michigan sport fishing generated more than $100 million. In comparison, even if all of Michigan's dormant hydro sites were renovated, the value of the electricity produced would be considerably less—approximately $10 million per year.

The fish passage issue in Michigan goes back to a law passed in 1929. The Fish Passage Act confers upon the Michigan Department of Natural Resources (DNR) the power and authority "to prohibit the obstruction of rivers, streams, and creeks in such manner as to prevent the free passage of fish up and down."

This law has seldom been used, but now, with renewed interest in both hydropower and inland sport fishing in Michigan, it has triggered a controversy. Part of the difficulty is the narrow mandate of some state agencies; the Michigan DNR in this case is specifically empowered to protect fish (as well as other natural resources), but not to promote power generation. To date, the

DNR has been lenient in enforcing the Act. It has, however, reserved the right to require fish passage retroactively at any site, and this could be very expensive for operators some time in the future. The issues surrounding the Fish Passage Act are just one example of regulatory complications that exist at the state level.

A Local Resource

Low-head hydropower has an important role to play, but that role should not be overestimated or developed at the expense of other natural resources. The old adage that "every little bit counts" certainly applies to energy production. Hydro redevelopment of any scale should be encouraged. Although its potentials are limited, hydropower has enough positive attributes to make it one of the most efficient and environmentally harmonious means of energy generation available today.

For hydro potential to be most appropriately assessed, it is best to view it as a *local* resource, not likely to contribute a significant amount of energy statewide. In Michigan, for example, the existing 80 hydroelectric sites generate about 2.7 percent of the state's total power. If the next best 30 sites were rehabilitated, they would not be able to push this figure to as much as 3 percent of Michigan's power needs.

A rural community, however, might find it more economical to augment its power needs by renovating a nearby hydro dam than by installing diesel power generators. Such local projects could be carried out in conjunction with park and recreational developments. Small industries could also take advantage of local hydropower potentials.

It would be a mistake simply to forget about the numerous decommissioned hydro dams. In many cases they have become unsightly rubbish heaps, or, worse, some are weakening structures that endanger life and property around them. The time has come for community groups, local governments, and other organizations to confront the problems and opportunities presented by these dams. While, in some cases, renovation for power generation by itself may not be economical, if these dams could be rebuilt to provide safe park facilities and local fishing sites, it is possible that such multiple uses of hydro sites could help pay for the renovations.

ACKNOWLEDGMENT

The authors would like to thank Lisa Holzman and Joe-Ellen Darch, both formerly of the MSU Department of Resource Development, for their helpful review of this article.

HYDRO: PAST OR FUTURE?[6]

During the 1970s, when energy resources seemed to be dangerously scarce and dauntingly expensive, Americans took a second look at thousands of small, abandoned dams spotted around the Northeast and Northwest. These dams had generated power early in the century, but fell idle when bigger, more centralized fossil-fueled generators took their place.

In 1978 the Public Utility Regulatory Policies Act (PURPA) made these old dams and powerhouses financially feasible again by making the utilities purchase their power. The rush to redevelop small-scale hydropower was sparked. Expectations were high in what industry folklore calls "the gold rush days."

Now the industry's outlook is rock bottom. Many observers predict the end of development of small-scale hydropower. But just as the early expectations were too high, today's gloom may be excessive, too.

Creating an Industry

PURPA encouraged the development of alternative energy resources to decrease U.S. dependence on imported oil. Besides forcing utility companies to buy power from small-scale producers at a fair price, the act exempted those producers from many utility regulations. Later, Congress added significant tax breaks for alternative energy producers, and high oil prices iced the cake. As a result, hydropower now generates close to 2,000 megawatts from nearly 200 sites, and the industry continues to grow.

[6]Reprint of an article by Sarah Cliffe, former editor of *Hydro Review*. Reprinted by permission from *Technology Review*, V. 89: 15, 18. Ag./S. '86. Copyright © 1986 by *Technology Review*.

Today, however, low oil prices make once-attractive hydro sites unworkable. Moreover, the tax breaks that sent entrepreneurs running to the nearest dam are all but gone. The energy tax credit for hydro expired at the end of 1985. Tax reform in Congress will probably take with it the investment tax credit and favorable depreciation rates.

Finally, environmental concerns, which once contributed to hydro's attraction, now threaten many projects. Excellent environmental arguments for development still exist: hydropower is clean, safe, renewable, and reliable. But problems have emerged. These include ensuring successful migration of fish around dam sites, keeping fish away from turbine blades, maintaining adequate instream flows to support recreation and animal life, and resistance from environmentalists to damming up undeveloped stretches of rivers.

In the 1970s, the energy crisis pushed regulators to find a middle ground between solving environmental problems and satisfying the need for renewable energy. The sense of crisis has since diminished, so regulatory agencies are less supportive of hydropower. In 1985 the Federal Energy Regulatory Commission (FERC), which licenses all power projects, changed the rules governing its licensing process. To be considered for an FERC license under the new rules, a developer must consult with—and have the approval of—local and state environmental regulators. In some states, this process is now well-nigh impossible to get through, even if a project is environmentally benign. Oregon is a prime example, having passed very tough anti-development laws in 1985.

Although some hydroelectric developers believe that the new FERC rules cede too much power to state agencies, many environmentalists accuse FERC of not being tough enough. In 1983 a coalition of environmental groups—including the Sierra Club, National Wildlife Federation, and National Audubon Society—asked the commission to give PURPA benefits only to existing dam sites. The coalition claimed that this would reflect the original legislative intent. When FERC did not respond, the coalition went to court to get an answer. The court in December 1985 gave FERC 90 days to act, and the agency reiterated its original position.

The coalition then requested that FERC do a major new study on the environmental impact of all hydro development. FERC answered that all potential sites are closely scrutinized for problems, that its Cumulative Impact Assessment Procedure monitors the effects of multiple-river basin sites, and that it already has a major impact study underway. The coalition considers these efforts insufficient and has gone back to court to force FERC to reopen the rules for granting PURPA benefits.

Congress Reconsiders

The ubiquitous description for the mid-1980s is "shakeout." The licenses for 65 existing projects, together generating about 700 megawatts, will expire in the next five years. Bills that establish standards for relicensing older dams have been passed by both houses of Congress and are currently in conference. Attached to these bills are provisions that would severely restrict the building of any new dams. The Senate version would allow new dams only on federally owned land, and only if they meet conditions set by fish and wildlife agencies. The House version puts a moratorium on all new dams until a study of PURPA benefits is completed.

In response to the more difficult market conditions, most developers who once embraced small-scale hydropower are gone from the scene. Those who remain approach the business with more sophistication. For example, Consolidated Hydro, Long Lake Energy Corp., and Catalyst Energy Development Corp. have made public stock offerings in the last six months. This move, unprecedented in small-scale hydro, puts their financing on a very different basis than relying on tax breaks.

The mom-and-pop operations are less in evidence; today's active developers can compete better because they own several sites and are looking for more. They will continue to consolidate, and some will continue to build, though the shrinking number of good sites will impose limitations. These survivors deal more effectively with environmental issues, choosing their projects judiciously and demonstrating both efficiency and good faith to regulators. By virtue of having developed several sites, these companies have established working relationships with agencies, bankers, insurers, and engineering firms.

In the long run, small-scale hydro's growth potential depends largely on when the energy shortage resurfaces, bringing high fossil-fuel prices. Public policy will no doubt reverse itself and reestablish the incentives that are all but gone today.

IV. SOLAR ENERGY, NUCLEAR FUSION, CONSERVATION

EDITOR'S INTRODUCTION

The final section of this compilation turns to alternative sources of energy such as wind, water, and solar energy, as well as technologies in an early experimental stage, such as power from nuclear fusion. Coastal regions of the country are now experimenting with ocean waves that, as they break on the shore, might be harnessed to generate a free and endlessly renewable source of electricity. In other regions windfarms (acres of blades rotated by the wind) have been constructed. But solar power, more than these others, has captured the public imagination; and its application in the Southwest particularly has fascinating possibilities. Nuclear fusion holds out another tantalizing prospect for the future, since it would vastly improve on nuclear fission and would produce little radioactive waste. But nuclear fusion is still in a theoretical stage, and its implementation, assuming that the technology can be perfected, would be far in the future. In another aspect of the fuel economy, the conservation of energy through such means as energy efficient homes and appliances has been notably successful in reducing energy consumption in the U. S. since the oil crisis of the 1970s, and it is commonly agreed that energy conservation should have a high priority in the time ahead.

The first articles in this section are devoted to solar energy. Writing in *Forbes,* James Cook calls attention to the work of Zoltan J. Kiss, Chairman of the Chronar Corporation in New Jersey, the country's largest and most successful producer of amorphous silicon photovoltaic cells that spontaneously transform light into electricity. As Cook notes, Chronar has cut production costs dramatically through new refinements of technology; if the efficiency of the photovoltaic panel can be improved further still solar energy may yet become competitive with the fossil-fuel industries. In a related article from *Science*, Robert Pool reports that Arco Solar Inc. has also scored a new breakthrough in thin-cell solar efficiency. In the larger context, however, solar energy is still a struggling young industry, overshadowed by oil and coal concerns, and as yet its contribution to the total energy supply is slight. The following staffwritten article from *Futurist* addresses the subject of nuclear fusion. The article points out that nuclear fusion would have significant advantages over nuclear fission, but that even if fusion reactors could be perfected and were economically feasible, their widespread use would not be likely to occur before the middle of the next century. A second article on the subject by Michael D. Lemonick, reprinted from *Time*, covers the most sensational and controversial news story on nuclear fusion of the decade—the announcement on April 24, 1989 that two University of Utah scientists, B. Stanley Pons and Martin Fleischmann, had achieved nuclear fusion in a jar at room temperature. If their claim can be validated, it would represent one of the major scientific developments of the century. But as the article points out, the nuclear fusion establishment in the U. S. has contested their claim vigorously, and at the moment of this writing the issue is still very much in doubt.

Not in doubt at all, however, are the advances made in energy conservation. Writing in *Sierra*, Seth Zuckerman cites the energy-saving steps made in the manufacture of automobiles and appliances, and in the construction of homes and public buildings; and predicts that conservation will be a permanent feature of our national energy policy. In a concluding article reprinted from *Science News*, Janet Raloff maintains that a conservation "revolution" is under way that will save hundreds of billions of dollars a year in fuel costs and reduce the level of pollution of the environment.

SOLAR ENERGY—GETTING HOTTER[1]

Zoltan J. Kiss is close to cashing in on a decade of breathtakingly hard work. A volatile and voluble man, Kiss is chairman of Lawrenceville, N.J.'s Chronar Corp. In Chronar, Kiss has put to-

[1]Reprint of an article by James Cook, *Forbes* staffwriter. Reprinted by permission from *Forbes*, V. 142:95–6. D. 12, '88. Copyright © 1988 by *Forbes*.

gether the U.S.' largest and most successful producer of amorphous silicon photovoltaic cells, those 1-by-3-foot panels that spontaneously transform light into electricity, and that for Chronar seem at last about to pay off.

Mail-order-catalog buffs are already familiar with at least one Chronar product, those solar-powered garden lights that require no wiring. But these are just frills. Chronar is into far more important stuff. Like winning a fair share of the world's future electricity market.

"To develop the technology for amorphous silicon," Kiss says, "we had to identify the manufacturing process, then build the equipment, because it didn't exist. Then raise the money and set up the factories. You couldn't have a product until you had the panels to put into product. One step after another. It took us a decade of incubation to get that far."

Photovoltaics long ago proved themselves in space, but costs were so high their commercial development has been slow in coming. All that is changing, thanks to the development of microscopically thin, amorphous noncrystalline coatings on glass, steel, aluminum or plastics, which cut production costs dramatically. With its costs continuing to decline, solar energy shapes up as the most benign answer to the world's growing energy and environmental problems.

There are other players in the photovoltaic business—Arco Solar, Energy Conversion Devices and Amoco's Solarex division, among others—but Chronar has gone further in bringing photovoltaics to the point of commercial development than any of its competitors. Either directly or through joint ventures abroad, Chronar can now manufacture 10,000 kilowatts of generating capacity a year and should have twice that next year.

Ten thousand kilowatts may not seem like much considering that the average new conventional power station runs around 600,000 kilowatts these days, but then Edison's original Pearl Street generating station started with only 360 kilowatts of capacity.

For Chronar, the explosion has already begun. Having diddled around at under $1 million in annual sales for nearly a decade, Chronar booked $11 million in 1986, $17 million in 1987, and should wind up with some $35 million this year and $70 million or more in 1989. "We're at the place where the company's sales are doubling," Kiss says, "and I think the doubling will continue at least for the next three or four years." That's a tall order. Keeping that up would make Chronar a half-billion-dollar-a-year company within a very few years.

Part of Chronar's growth has come from sales to affiliates of equipment for manufacturing solar panels. But Chronar is manufacturing more and more end products: Besides the walkway and garden lights for the consumer market, there are billboard lights, highway lighting and water pumps for industrial markets, especially in areas beyond the reach of conventional power sources. Last year Chronar sold only $4 million worth of consumer products. This year it will sell about $13 million. Last year it sold no industrial products. This year it will sell over $1 million. Next year, depending on the rate of market development, it should sell $10 million or more.

Best of all, after piling up more than $20 million in losses in the past five years, Chronar will probably break even this year though nine-month earnings were still $1.1 million in the red. "For ten years we were in the development stage," Kiss says. "We are now at the point of turning into the black, and from now on we expect to be profitable."

A physicist by training, Kiss left his native Hungary in 1950 at the age of 17, picked up a Ph.D. at the University of Toronto, and wound up in the early 1960s as head of quantum electronics research at RCA Laboratories in Princeton, N.J. He quit RCA Labs in 1969 to start a liquid crystal display outfit called Optel, quit again in 1976 when Mitsubishi gained control of the company. He went out on his own, choosing photovoltaic technology and putting up $40,000 of his own money to found Chronar in 1976.

Half the battle has been finding the money to finance the losses inevitable in any high-tech development business, and it didn't help any that a few years ago the SEC went after Chronar and its accountants for what it considered fraudulent accounting practices. That stalled Chronar for a time, forcing the cancellation of a $12 million public stock offering. Kiss eased his financial problems by setting up joint-venture manufacturing partnerships around the world and in the last year or two has picked up a handful of well-heeled backers. These include Birmingham construction tycoons John and Bill Harbert (who made a $5 million equity investment), the Sheet Metal Workers' pension fund ($7.5 million in equity, $7.5 million in loans), the National Electrical Contractors' pension fund ($10 million in equity) and most recently Pacific Gas & Electric ($1.8 million).

Chronar can already compete, Kiss claims, with small-scale diesel generators ranging up to 50 kilowatts, producing electricity for 25 to 35 cents per kilowatt-hour, versus 30 to 50 cents for the diesels, depending on their size and other factors. And that, says Kiss, opens up a $5 billion market. But the really big market is yet to be conquered—the one served by large-scale coal- or nuclear-fueled systems, which retail power at under 7 cents a kilowatt-hour. If photovoltaics can crack that, the potential is incalculable but surely vast. To judge by the huge short interest in Chronar's stock, not everyone agrees.

As Kiss likes to point out, photovoltaics is a modular technology. To increase capacity, you don't build larger plants, you just keep adding panels and interconnecting them. You don't have to hook into the utility grid. Photovoltaics can operate independently and effectively on their own. Chronar itself is putting a 60-kilowatt station on the roof of its New Jersey headquarters, which will generate power for about 20 cents a kilowatt, comparable to the 19 cents the local utility, Public Service Electric & Gas, charges at its summer peaks.

To crack the big-scale utility market, Chronar will have to do a lot better than that. The principal cost in photovoltaic power generation is the cost of the capital needed to produce and install the photovoltaic panels, and though Chronar has reduced such costs by a third in the past 18 months, it's far from enough. Chronar has recently worked out a $25 million joint venture with Bechtel, a Pacific Gas & Electric subsidiary and a Sheet Metal Workers' pension fund subsidiary to build an automated manufacturing plant, slated to go into operation next year, that is expected to cut the direct cost of manufacture by 50%. At that level, Chronar can install a power station for $2,500 a kilowatt and make a decent profit. Says Kiss: "We will have brought the cost of manufacturing down to the point where we can generate electricity in the 10-to-12-cents-a-kilowatt-hour range." That's still not enough: The utility industry generates power at a fully loaded average cost of approximately 6 cents per kilowatt-hour.

Nevertheless, Kiss is so confident that the economics of the new plant will prove out that he is working out another joint venture with SeaWest Industries, a California wind power company, to build a proposed $125 million, 50,000-kilowatt photovoltaic plant that will sell power to Southern California Edison during peak periods. Under legislation designed to encourage the use of alternative energy sources, the rates will range up to 11 cents a kilowatt-hour in 1989 and up to 24 cents in the year 2000.

At such prices, Chronar begins to move in on its utility targets, and Kiss is confident that within five years he'll be able to reduce costs another 50% by doubling the efficiency of the photovoltaic panel. If he succeeds—and that may not be so easily done as he claims—Chronar will be able to generate power for under 7 cents a kilowatt-hour, and the long-awaited breakthrough will be achieved.

It may not happen, but neither is it implausible. San Francisco's Pacific Gas & Electric is persuaded enough that it bought a 250,000-share stake in the company. "If solar panels make the economic breakthrough necessary for them to become marketable," says Richard Clarke, chairman of the big utility, "I want PG&E people installing them for our customers."

Clarke says "if," but for Kiss there are no "ifs." He foresees the day when photovoltaics will be used in the electrolysis of water to produce cheap hydrogen for fuel cells and combustion engines, opening up to photovoltaics not just the market for electricity but also the transportation market. Eat your heart out, OPEC.

Kiss becomes almost lyrical when talking of the future. Says he: "I believe photovoltaics will ultimately be the world's primary energy source, and it's just a question of how rapidly the cost can be brought down."

A BRIGHT SPOT ON THE SOLAR SCENE[2]

Arco Solar Inc. recently announced what some are calling one of the biggest breakthroughs in the history of solar energy. The company, a unit of Atlantic Richfield Corp., has developed a thin-film solar cell with an 11.2% efficiency—much higher than

[2]Reprint of an article by Robert Pool, *Science* staffwriter. Reprinted by permission from *Science*, V. 241:901. Ag. 19, '88. Copyright © 1988 by the American Association for the Advancement of Science.

existing cells of similar type. If the new solar cell lives up to its billing, it just might make collecting sunlight a commercially feasible alternative to smashing atoms or burning fossil fuels.

"It's a real landmark, the biggest news to break this year in solar cell research," said Jack Stone, director of solar electric research at the Solar Energy Research Institute (SERI), which worked with Arco Solar on the new cell. SERI, which is owned by the Department of Energy, is spending $4.2 million and Arco Solar is spending $4.8 million in a 3-year partnership to develop solar cells and panels.

The reason for the excitement is that Arco Solar and SERI have taken a giant step in an industry where baby steps are the norm. In the struggle to squeeze more electricity out of sunlight, it is considered a victory to raise the sunlight-to-electricity conversion ratio by a few tenths of a percent. Yet Arco Solar has, in one step, developed a solar cell whose working efficiency is a good three percentage points higher than the best the solar industry could offer just a few months ago.

The breakthrough came in the area of thin-film solar cells, a low-cost type of cell whose low efficiency is offset by the fact that it is relatively inexpensive to manufacture. (A thin-film cell, like it sounds, consists of a film of silicon or other photosensitive substance deposited on a base layer, as opposed to a single crystal, which is one solid piece of silicon or other material.) Solar-powered calculators, for example, use thin-film solar cells with sunlight-to-electricity conversion ratios of 3% or less. Although thin-film cells are much less efficient than single-crystal cells—the best of which now reach 28 to 29% efficiency—the fact that thin-film cells can be made cheaply makes them the favorite candidate to produce commercially competitive solar power for utilities.

The giant step in efficiency came from using a relatively new material for the cell. Most thin-film cells are made of silicon, and the best large silicon thin-film cells have efficiencies of just over 9%. (Smaller cells can have larger efficiencies because it is easier to make a small cell more uniform, but the industry judges cells for use in generating electric power on the basis of how efficient a 1-square-foot cell is.)

Arco Solar reached 11.2% efficiency in a 1-square-foot copper-indium-diselenide (CIS) cell. Charles Gay, senior vice president of manufacturing, research, and engineering at Arco Solar, said the breakthrough came from "a number of years of hammering away at a number of problems." As early as late 1985, he said, his lab had a small CIS cell with 12.5% efficiency, but scaling that up to a practical size took nearly 3 years.

The jump from 9 to 11.2% is significant by itself, but the CIS cell has another big plus. Silicon cells degrade when exposed to sunlight and quickly lose 10 to 20% of their efficiency, but the CIS cell seems to remain stable, according to Ken Zweibel at SERI. This means that the best available thin-film silicon cells can be expected to perform at 8% efficiency or less, as compared to a stable 11.2% for CIS.

The CIS solar cell has five layers. The substrate is topped with a layer of molybdenum that serves as an electrical conduction layer for the back of the cell. Then comes a 1-micrometer layer of copper-indium-diselenide ($CuInSe_2$) and a 0.03-micrometer film of cadmium sulfide (CdS). On top is a layer of zinc oxide (ZnO) that is transparent to sunlight, so that light can pass through to the CdS and $CuInSe_2$ layers, but that is electrically conductive to serve as the second electrical contact layer.

Once a plant is set up to manufacture CIS cells, the cells should cost about $100 a square meter to make, or about the cost of the existing commercial thin-film silicon cells with 7% efficiency. Ultimately, the goal is to bring the manufacturing costs down to $50 a square meter and improve the efficiency to 15%, at which point the Department of Energy estimates solar energy could be competitive with other forms of power.

For now, Gay said, Arco Solar will concentrate on turning a laboratory product into a commercial one. The company will need to start making large quantities to see what the actual cost is, and will provide the CIS cells to SERI and utility companies for testing, he said. "It takes a whole lot of work to bring this to market."

FUSION: FUEL FOR THE FUTURE?[3]

Nuclear fusion could provide humanity with a virtually unlimited source of safe electricity, but probably not in the near future, according to the U.S. Office of Technology Assessment (OTA).

Fusion could be an attractive source of electricity because it avoids the environmental and safety problems of fossil fuels and nuclear fission. Unlike fossil fuel power plants, fusion reactors should not produce air pollutants, such as carbon dioxide—a greenhouse gas. And unlike fission, the process currently used in nuclear power plants, fusion would not produce high-level, long-lived radioactive waste, says OTA in a new report, *Starpower: The U.S. and the International Quest for Fusion Energy.*

Fusion research and development has been ongoing in the United States since the 1950s, with federal funding growing tenfold in the 1970s. But the U.S. budget deficit, coupled with a reduced sense of urgency in finding alternatives to coal and oil, has resulted in cuts in the U.S. fusion program budget to about one-half of its peak level a decade ago, says OTA.

Fusion of hydrogen nuclei requires heat of about 100 million degrees Celsius. At that temperature, matter exists as plasma; but no solid container has been developed that can withstand the heat. One potential solution is magnetic fields to confine the plasma.

The essential problem for scientists and engineers to overcome is bringing fusion reactions to the break-even point, where as much energy is produced by the reaction as is required to maintain it. This milestone may be reached by 1990, says OTA. High-energy output—beyond the break-even point—will be a more difficult task, but essential to demonstrate a commercially viable future for fusion.

To prove economic viability for fusion energy, a prototype commercial reactor would need to be developed. But at least three more decades of R&D are required before such a prototype could be operated and evaluated, OTA estimates. Widespread

[3]Reprint of a staffwritten article in *Futurist.* Reprinted by permission from *The Futurist,* V. 22:2. M./Ap. '88. Copyright © 1988 by World Future Society, 4916 Saint Elmo Avenue, Bethesda, Maryland 20814.

use of fusion energy would take several more decades beyond that.

The U.S. Department of Energy hopes to develop a prototype reactor by the early part of the twenty-first century, but the current U.S. fusion budget falls short of what is needed to meet that goal independently, says OTA. International cooperation in fusion R&D has a long and successful history, but the domestic program should not be sacrificed in favor of international collaboration, the report warns.

FUSION ILLUSION?[4]

Little more than a month ago, they were just two chemists, toiling in virtual anonymity. But B. Stanley Pons and Martin Fleischmann came last week to Washington as heroes, visionaries and scientific superstars. With a mob of reporters following along, the thermodynamic duo marched onto Capitol Hill to tell Congress how their simple tabletop experiment had generated fusion, the nuclear reaction that powers the sun. Displaying slides filled with complex equations, wielding electronic pointers and pulling a mockup of their apparatus from a plastic shopping bag, the bespectacled researchers mesmerized the members of the House Committee on Science, Space and Technology with an account of how their device produced more energy, in the form of heat, than it consumed. The politicians may have been baffled by the chemistry, but they had no trouble grasping the implications. It seemed that Pons, a professor at the University of Utah, and Fleischmann, of Britain's University of Southampton, might have pulled off a trick that has eluded some of the best minds in physics for nearly four decades. More important, they might have found a way to solve the world's energy problems for all time.

What would it take, they were asked, to make that dream a reality? Money from Congress, of course. University of Utah President Chase Peterson, who was right there at the scientists' side, suggested that $25 million would be a nice sum to help his

[4]Reprint of an article by Michael D. Lemonick, *Time* staffwriter. Reprinted by permission from *Time*, May 8, 1989, pp. 72-77. Copyright © 1989 by Time Inc.

school set up a fusion research center. Some of the Congressmen appeared eager to oblige. "Today," rhapsodized Robert Roe, a New Jersey Democrat, "we may be poised on the threshold of a new era. It is possible that we may be witnessing the cold-fusion revolution."

But Congress had better wait a while before it starts pouring taxpayers' money into Utah's test tubes. Even as Pons and Fleischmann stirred excitement on Capitol Hill, evidence was mounting that their form of fusion is probably an illusion. More and more scientists were openly scoffing at the chemists' claim that they had caused deuterium ions, which are commonly found in seawater, to fuse to form helium, liberating large amounts of heat. Physicists have never been able to achieve such a sustained reaction, even briefly, without subjecting deuterium to the kind of extreme temperature and pressure found inside the sun.

While no one has proved conclusively that Pons and Fleischmann are wrong, it seems likely that they jumped to a hasty conclusion based on incomplete research. Scientists in Japan and Switzerland announced that their own tests had convinced them the original work was flawed. An attempt by the Harwell Laboratory in Britain to confirm the discovery has also produced nothing, even though Fleischmann himself checked the experiments.

None of the major national laboratories in the U.S. have obtained positive results either. This week data from one of the most comprehensive sets of experiments to date—a collaboration between Brookhaven National Laboratory and Yale University—will be presented at the spring meeting of the American Physical Society in Baltimore. The Brookhaven-Yale tests found no evidence of what Pons and Fleischmann saw. Brookhaven physicist Kelvin Lynn speculates that the heat produced may possibly be the result of some more conventional, though unexpected, chemical reaction. "It's quite interesting," he says, "to wonder how nature may have conspired to make them believe they had fusion."

Most damning of all is the editorial that appears in the current issue of the prestigious British journal *Nature*. The Pons-Fleischmann claim, writes editor John Maddox, "is literally unsupported by the evidence, could be an artifact [a spurious result unrelated to the phenomenon under investigation] and, given its improbability, is most likely to be one." Maddox noted that the team announced its results before performing even the most basic control experiments to verify the findings. That was an "astonishing oversight," wrote Maddox, "a glaring lapse from accepted practice."

Those are strong words, but Pons and Fleischmann are hanging tough behind their claim. Pons, in fact, says the experiments in his Utah lab have begun to produce increasing amounts of heat. And he has picked up a determined band of supporters. Robert Huggins, a respected materials scientist at Stanford, contends that he has also obtained excess heat in a series of similar experiments. Says Huggins: "The magnitudes of our observed effects are comparable to those reported earlier by Fleischmann and Pons, and lend strong support to the validity of their results."

Whether or not they turn out to be right, Pons and Fleischmann have pushed the entire scientific world into a frenzy. After the March 23 press conference in which the two chemists went public with their discovery, researchers around the globe immediately came down with fusion fever. Its symptoms were hyperactivity, insomnia and delusions of grandeur. Gleaning what meager information they could from murky faxes of an unpublished Pons-Fleischmann paper and from TV pictures of apparatus, chemists and physicists dropped whatever else they were doing in attempts to verify or shoot down the concept of cold fusion.

Thus began one of the strangest months in the history of science. Hardly a day passed without an announcement from somewhere—Texas, Georgia, Hungary, Brazil, India, the Soviet Union—that at least some parts of the Pons-Fleischmann experiment had been replicated. Scientific protocol went out the window as researchers called press conferences to trumpet the latest results before verifying them.

That turned out to be a dangerous course. The Georgia Institute of Technology, for example, claimed that its team had detected neutrons, a hallmark of fusion reactions, coming from a setup similar to the one Pons and Fleischmann had used. But then the scientists had to retract the assertion, admitting with embarrassment that they had been misled by a faulty neutron detector. And chemists at Texas A&M, who initially reported significant amounts of excess heat generated by their device, were disappointed when they got less heat in later experiments.

This new phenomenon of science by press conference disturbed many researchers. Said Moshe Gai, a Yale physicist and a member of the Yale-Brookhaven collaboration: "I am dissatisfied and somewhat disappointed with some of my fellow scientists who have done things too much in a hurry." Charles C. Baker, director of fusion research at Argonne National Laboratory, was blunter: "Calling press conferences and making claims of results without having a well-prepared technical report is not the way for a good, professional scientist to function."

Equally offensive to many scientists is the fact that Pons and Fleischmann have steadfastly refused to disclose important details of their work that would enable others to duplicate it. Though they eventually published an account of their experiments in the *Journal of Electroanalytical Chemistry and Interfacial Electrochemistry*, a highly technical Swiss periodical, the paper was too sketchy to be truly enlightening. Pons has argued repeatedly that his critics who are getting negative results do not know how to run the experiment, but he does not show them precisely what they are doing wrong. Declares Keith Thomassen, a physicist who heads one of the fusion-research programs at Lawrence Livermore National Laboratory: "The hard, uncompromising way in which we do our business is that when you make a claim, you present the facts on which you base that claim."

Why is Pons being so cagey? Perhaps because the discovery he and Fleischmann claim to have made could be worth a fortune. Keeping some of the secrets to themselves could serve to protect their financial interests and those of the University of Utah, which has already filed five patent applications, with more to come. Pons insists, though, that he has reached an agreement with Los Alamos National Laboratory to help its scientists replicate his cold-fusion experiments.

The awesome potential of the alleged discovery explains why so many people are badgering Pons and Fleischmann for information, and why they are giving it out so cautiously. A practical technique for creating useful fusion energy at low temperatures could change the world forever by providing a source of virtually limitless power. Moreover, the process would generate no pollutants—not even carbon dioxide, which many scientists fear is warming the globe in a greenhouse effect. A fusion plant would give off much less radiation than do conventional nuclear-power generators. And it would essentially run on seawater. Any scientist who managed to harness fusion would be guaranteed a Nobel

Prize for Physics (and probably Peace as well), untold riches from licensing the process and a place in history alongside Einstein and slightly above Edison. Any scientist who confirmed the claim would get part of the resulting avalanche of research dollars, and anyone who shot it down would gain acclaim within the scientific community.

But the reasons for the fusion furor are more complicated than just the prospects of riches and fame. Scientists and university administrators are sometimes driven by the same sort of base emotions—like jealousy and paranoia—that often motivate less intellectually lofty folks, and the peculiar circumstances of this discovery helped ignite a number of long-smoldering resentments. For one thing, fusion and other subatomic phenomena that are usually studied with giant nuclear reactors and particle accelerators have long been the private domain of physicists. Chemists, on the other hand, were more likely to be studying how to make a better laundry detergent, or so physicists seem to think. It is no surprise, then, that the harshest critics of Pons and his dime-store equipment have been physicists. Retorts Pons: "Chemists are supposed to discover new chemicals. The physicists don't like it when they discover new physicals." In fact, many chemists feel—with much justification—that the physicists consider themselves intellectually superior. Says Cheves Walling, a Utah chemist who has developed one theory to explain how the cold-fusion experiment might work: "Chemists resent the fact that physicists can get money for multimillion-dollar experiments that could have gone to chemists to do something more useful."

Still, the cold-fusion combat is not just the physicists vs. the chemists. There is a sense in Salt Lake City that most of Pons' critics are what Utah chemist David Grant calls "the mean bullies from the Eastern establishment." Such snotty folks should remember, he says, that "science is not the domain of one set of colleges or one set of people anymore."

There is also an intense rivalry between the University of Utah—the U, for short—and Brigham Young University, located just 50 miles away. Although the U is the state-supported university, Utah's majority Mormon population identifies far more strongly with church-run Brigham Young. It was at least partly because a Brigham Young physicist named Steven Jones was nearing an announcement on cold fusion too that Pons and

Fleischmann called their surprise press conference. They had been urged to go public by University of Utah administrators, who were apparently fearful that archrivals at Brigham Young would steal the fusion spotlight. The U has had chronic money troubles recently, and an influx of fusion-research grants, not to mention international glory, could go a long way toward remedying the situation.

Any serious prospect of practical fusion will attract federal research funding. For decades the Government has spent billions of dollars in pursuit of this tantalizing but elusive goal. The first man-made fusion reactions took the form of H-bomb explosions in the 1950s. Scientists then set out to bring that incredible power under control. Their strategy was to confine deuterium, a heavy form of hydrogen, within a "bottle" of magnetic force and heat it to tens of millions of degrees. The nuclei of the atoms, forced close together despite their mutually repellent positive electric charges, would fuse, releasing energy. Elaborated and modified, that is the approach still being taken at such state-of-the-art facilities as Princeton's Plasma Physics Laboratory. But the lab has achieved only brief bursts of fusion at enormous cost. A more recent concept, represented by Livermore's Nova machine, is to take tiny "marbles" filled with deuterium and concentrate 100 trillion watts of laser light on them for a billionth of a second. The deuterium should theoretically fuse and produce energy, but a far more powerful laser would be needed to spark a useful reaction.

Although superhot fusion has always been considered the best way to generate power, physicists have known since the 1950s that the process can take place at room temperature as well. If the electrons in deuterium are replaced with heavier particles called muons, the deuterium nuclei can approach each other more closely and occasionally fuse on their own. This muon-catalyzed fusion has never produced significant amounts of energy.

Yet the thought that cold fusion was possible at all continued to intrigue some scientists, including Pons and Fleischmann. When Pons got his Ph.D. at Southampton in 1978, Fleischmann was his department head. They became close friends and collaborators after Pons graduated, and remained so when he settled at Utah. One day in 1984 Pons and Fleischmann had a sudden idea for a new way to achieve cold fusion. The brainstorm came, Pons says, during a hike up Millcreek Canyon, near his home in Salt Lake City. He and Fleischmann were puzzling over the peculiar properties of certain metals, like palladium, that are known to absorb huge quantities of hydrogen gas. In the presence of an electric field, the chemists had noticed, deuterium nuclei appeared to be unusually free to move around within palladium's latticework of atoms. They speculated that the nuclei might even come close enough together to make nuclear fusion more likely.

"We came down from the hike," recalls Pons, "and then we stood around the table in my kitchen, had a couple of Jack Daniel's and started drawing pictures." Their experiments resembled nothing more than the simple electrochemical cells often entered in high school science fairs: two metal electrodes immersed in a bath of water laced with mineral salts and connected to a power supply. The only differences were that one of the electrodes was made of palladium and the water was heavy water, or deuterium oxide (chemical formula D_2O), rather than ordinary H_2O.

The first experiments did not do much. But one night in 1985, an electrochemical cell being used by the two scientists melted down. "That," says Pons, "told us we had much more energy than could be attributed to a chemical reaction." After the accident, Pons called Fleischmann, who had returned to England. Fleischmann responded to the momentous news with an admonition: "We'd better not talk on the phone." Pons says they ultimately spent about $100,000 of their own money to pursue what they were convinced was fusion.

Neither Pons nor Fleischmann would have ranked high on anyone's list of scientists likely to revolutionize physics, although both are respected researchers in the field of electrochemistry, the study of how chemical reactions behave in the presence of an electric field. In retrospect, though, their backgrounds were quirky enough to suggest that almost anything was possible. Pons, in particular, had an unorthodox professional history. A native North Carolinian, Pons, 46, dropped out of graduate school at the University of Michigan in 1967, just a few months shy of getting a Ph.D. in chemistry. "Jobs for Ph.D. chemists were paying $3,500 a year at the time," he explains. "My daddy offered me $20,000." He joined the family textile business, then went on to manage a family-owned restaurant in North Palm Beach, Fla. But after nearly a decade away from science, Pons decided to go back and complete his degree. To do so at Michigan, though, he would have had to repeat most of the courses he had already taken. So he went to Southampton, where his credits would still count.

After finishing his doctorate, Pons was able to make up for lost time, becoming chairman of the Utah chemistry department in 1988. Along the way he earned a reputation for diligence and creativity. Says Harry Mark, Pons' adviser at Michigan: "Stan was innovative and controversial even back in grad school. What he's doing now doesn't surprise me."

Fleischmann too is known for resourcefulness. Now 62, he arrived in England in 1939 with his family, Czech refugees from Hitler's Europe, and soon distinguished himself in school and college. Jan Fells, who worked with him at the University of Newcastle, calls him a man of "great ideas," and Roger Parsons, head of the chemistry department at Southampton, describes Fleischmann as "excitable in the sense that he gets very enthusiastic about ideas. He is a man full of ideas across a wide field and not necessarily connected to his main research."

By 1988, Pons and Fleischmann were focusing much of their attention on the quest for cold fusion. But they were not alone. At Brigham Young, a team headed by physicist Steven Jones had been working on a similar experiment for at least two years. Jones had also found evidence of fusion, but did not get the excess heat production that Pons and Fleischmann were observing. The two groups were evidently unaware of each other until last September, when Jones was asked to review a Pons-Fleischmann grant application. To his surprise, Jones says, he realized that he and the Utah researchers were following parallel paths. He made contact with Pons and suggested that the unwitting competitors should collaborate.

That eventually led to a showdown meeting on March 6 at which, according to a Brigham Young document, the scientists and top administrators from both universities were present. At issue was the timing of public statements. Pons and Fleischmann said they would prefer to wait before releasing results. Jones countered that he had been invited to talk about his work before the American Physical Society in May and that he intended to do so. According to Brigham Young, the meeting ended with an agreement to submit simultaneous papers to *Nature* on March 24. When Pons and Fleischmann suddenly announced their "breakthrough" on March 23, Jones felt he had been sandbagged.

The race with Jones appears to have forced Pons and Fleischmann to go public long before they were ready. Their paper on cold fusion is considered less—far less—than rigorous. "Every great discovery has had plenty of skeptics," notes Richard Muller, a physicist at Lawrence Berkeley Laboratory, "but I can't find any great discovery of the past 50 years that was published with a bad paper. If a freshman physics or chemistry major had done it, they would have flunked." Says Robert G. Sachs, former director of Argonne National Laboratory: "It doesn't meet the kind of standards you'd want to meet for nuclear physics. It doesn't even meet the standards of testing in inorganic chemistry. It's a shame. They obviously just got too excited about it to think straight."

Nature asked for more information from Pons and Fleischmann before publishing the paper, but according to the journal the pair said they were too busy. Fleischmann, though, claims they supplied 19 new pages. In any case, the paper was withdrawn. Says Fleischmann: "*Nature* is not the appropriate place to publish because they don't publish full papers." That peculiar sentiment might come as a surprise to James Watson and Francis Crick, whose Nobel-prizewinning discovery of the structure of DNA was first published in the British journal.

None of the criticisms leveled at Pons and Fleischmann mean that they are necessarily wrong. But the burden of proof remains on them. So far, they have failed to demonstrate convincingly that they have indeed produced a new sort of fusion. And if the two chemists cannot think of any way to explain the excess heat in their experiment without resorting to nuclear reactions, others can. Chemist Linus Pauling, a Nobel laureate and himself something of an iconoclast, thinks that when absorbing high concentrations of deuterium, the palladium lattice may become unstable and deteriorate, releasing heat.

Even if Pons and Fleischmann should turn out to be right, the world's energy problems are not necessarily over. As the proponents of more conventional fusion research have learned, transforming a reaction from a laboratory curiosity to a full-scale energy technology can be incredibly difficult. Magnetic fusion has yet to achieve break-even, the stage at which the amount of energy coming out is equal to that going in. Says Harold Furth, director of Princeton's effort: "We are essentially within a factor of two of break-even now. Seeing that it used to be a factor of a million, we feel extremely optimistic." But it has taken more than 30 years to get there, and plenty of technical problems remain.

In short, no matter which scheme proves best, the virtually limitless power that could eventually result from fusion is a dream that will not come true anytime soon. The solution to the world's energy crisis is not likely to be declared in a press conference. It must be slowly and carefully worked out, step by painstaking step.

WHAT EVER HAPPENED TO ENERGY CONSERVATION?[5]

When the end came, it came with the suddenness of an oil gusher. Last December, OPEC oil ministers once again failed to agree on keeping oil flows down and prices high. But this time the Saudis let the bottom fall out of the market. Crude oil, which had slid from $35 a barrel in 1981 to $26 by the end of 1985, plunged to $10 on the world market in April. Gasoline prices started a parallel drop, diving to levels unknown since the Iranian oil shock of 1979.

Will this price crash—an obvious boon to consumers—spell an end to conservation and efficiency improvements? The picture is not as gloomy as it might appear. Analysts predict that the drop in oil prices will slow the transition to more efficient use of energy, but they also agree that the collapse of OPEC does not augur a return to the wasteful days of the 1960s, when Americans were running through natural resources at blinding speed. "Energy efficiency is to a large extent entrenched," says Howard Geller, associate director of the American Council for an Energy-Efficient Economy. "Things will continue to improve, but the rate at which they get better may decrease."

Analysts say advances in automobile fuel economy will be delayed the most, while the move toward more efficient use of electricity is likely to remain relatively unaffected, despite the surplus of generating capacity that has appeared over the last three years. From energy codes to fuel economy standards, energy efficiency now has institutional momentum behind it. That momentum will

[5]Reprint of an article by Seth Zuckerman, author and energy policy specialist. Reprinted by permission from *Sierra*, V. 71:58–62. S./O. '86. Copyright © 1986 by *Sierra*.

be crucial in averting the risks of renewed dependency on oil—which would lead to another energy crunch in the 1990s.

The recent collapse of oil prices follows a decade of increasingly efficient use of energy. The United States' dependence on energy, measured by the amount required to produce a real dollar of GNP, has dropped by more than 25 percent since 1973. Researchers at Lawrence Berkeley Laboratory in California attribute about half the drop to "technical fixes"—changes that enable people to perform the same services using less energy. Some examples: The average car goes 29 percent farther on a gallon of gas now than it did in 1973; millions of homes have been insulated; factory heat that was once wasted is being recovered; and new refrigerators and air conditioners are more efficient than the models they replaced. The rest of the drop has resulted from consumer belt-tightening, the transition from a manufacturing to a service economy, and changes in consumer preferences.

Both government regulation and economic forces have spurred these changes. Congress set the Corporate Average Fuel Economy (CAFE) standards in 1975, mandating that each manufacturer's new fleet average 27.5 miles per gallon by 1985. Although the standard was rolled back to 26 mpg in 1985 as a special favor to Ford and General Motors, it played a major role in boosting the typical car's 1974 average of 13 mpg.

Some states and many cities have passed codes that set limits on the energy consumption of new buildings. Several large states have led the way with appliance efficiency standards. For example, California plans to cut the maximum permissible electricity use of an average-size refrigerator from 1,400 to 700 kilowatt hours per year by 1992. Because California represents about a tenth of the U.S. appliance market, this change will have an effect on manufacturers' product lines nationwide.

Even if the oil crash keeps standards from being stiffened further, analysts agree that they will probably not be weakened, and will continue to pull up overall energy efficiency.

Government regulation isn't the whole story, however: Price hikes have also led people to use energy more efficiently. Between 1978 and 1981, oil prices doubled and gasoline prices leaped 60 percent above the rate of inflation. Electricity prices rose by a fifth between 1978 and 1982. Energy analyst Amory Lovins estimates that about a third of the country continues to experience electric "rate shock"—sudden, sizeable increases in rates, usually caused by new facilities.

People have responded by buying devices that provide the services they want while using less energy. Manufacturers now produce refrigerators and air conditioners that are more efficient than required by the strictest standards. Numerous energy-saving products are succeeding in the marketplace even though their installation is not mandated by state or local codes. Some auto manufacturers exceed the CAFE standards handsomely.

Many of these changes, especially fuel-frugal cars (along with the plentiful supply of natural gas as a substitute for oil), contributed to the oil glut. Between 1979 and 1984, oil consumption fell by one sixth in the industrialized Western world. If the United States still used the same amount of oil per unit of real GNP that it did in 1978, it would be importing nearly three times as much oil as it is today.

While demand dropped, non-OPEC oil exporters such as Mexico, Great Britain, and Norway increased oil production. As a result, OPEC's share of the world market dropped from 42 percent in 1979 to 30 percent in 1984, and its revenues from $275 billion in 1981 to $150 billion in 1984, with a drop to $100 billion likely in 1986 if current trends continue.

Analysts agree that oil prices are likely to start climbing in the late 1980s or '90s, but are uncertain exactly when. "Anyone who tries to sell you a price-and-demand forecast is dealing more with hot air than anything else," says Worldwatch Institute senior researcher Christopher Flavin.

But analysts do agree on certain trends. The Mideast members of OPEC hold 56 percent of the world's oil reserves. "At some point, dependence will shift back to the Middle East," says Daniel Yergin, co-author of the Harvard Business School's *Energy Future.* Flavin points out that OPEC members such as Algeria, Ecuador, Indonesia, and Nigeria are likely to exhaust their exportable reserves by the mid-1990s, making the cartel "smaller, more geographically concentrated, more cohesive, and more powerful." American oil consumers face the danger of being fooled by the cycle of crisis and glut into letting energy efficiency decline.

Whatever effect the price crash has on oil use, it will have much less of an impact on the consumption of other fuels. A 1986 report by the American Council for an Energy-Efficient Economy shows that oil represents less than 3 percent of the cost of generating electricity nationwide. Observers suggest that the oil price drop will therefore have little or no effect on the retail price of electricity. On the other hand, it is theoretically possible to substitute oil for natural gas, driving down the price of gas. But Lovins plays down this possibility, saying that such substitutions are no longer as feasible technically as they used to be.

Transportation still depends on petroleum; here the effects of the OPEC collapse are as plain as the price signs at the corner gas station. Yet, even in transportation, analysts believe progress in energy efficiency will be slowed but not stalled. "Detroit has re-tooled," says Yergin. "It's awfully hard to buy a gas-guzzler—unless you can afford a Rolls-Royce."

Indeed, the average new domestic car gets more than 26 miles to the gallon, a substantial improvement over the 17 mpg of the average car on the road, and an even greater leap from the mileage of the cars being retired from the road. The new American cars being made by foreign manufacturers (35 percent of the total) are even more impressive: They average more than 30 mpg.

One possible blow to fuel economy would be growth in the size of cars that people buy. Sierra Club Washington lobbyist Brooks Yeager sees a move toward larger and less efficient cars. The oil crash, he says, "allows consumers to believe they will be able to afford the gas." But there's a limit to that trend. "It's not as though we're going back to 1965 cars," says Lovins, and even today's large cars are about as efficient as some ten-year-old subcompacts.

The drop in oil prices has made a rollback of the CAFE standards a near certainty, however, as demonstrated by the National Highway Traffic and Safety Administration's lowering last year of automakers' required average mileage for their fleets from 27.5 to 26 mpg. No one in Congress could muster the votes necessary to stop the rollback, much less raise the standards to 45 mpg by 1995, as Reps. Barbara Boxer (D-Calif.) and Claudine Schneider (R-R.I) proposed.

It's now clear that both the administration and Congress are willing to keep the standard at 26 mpg through 1988, Yeager says. "There's a lost sense of urgency about forcing the U.S. auto industry to conserve energy."

Although U.S. auto manufacturers may lag behind, more efficient cars continue to appear on showroom floors. One is the imported Chevrolet Sprint, which boasts an EPA rating of 55 mpg in the city and 60 on the highway. "The top-rated cars still improve from year to year," says Geller.

Deborah Bleviss of the Federation of American Scientists cites prototypes by Volvo (63 mpg city, 81 highway) and Toyota (89 mpg city, 110 highway) as indications of how far automakers could go. To achieve mileage increases, these cars use plastics and ceramics in auto engines and plastics in auto bodies, as well as even newer technologies such as direct-injection diesel engines.

Outside the transportation sector, the prospect for increased energy efficiency continues to look positive despite the dive in oil prices. Homeowners are likely to continue to seek efficiency improvements, says Lovins, "because practically all home energy fixes are cost-effective down to $5 a barrel." A 1985 survey by the National Association of Home Builders found that homebuyers mentioned energy efficiency as a priority in a new home more often than any other factor—somewhat less often than in a 1980 survey, but far ahead of the next-most-important factors, less exterior maintenance and a newer house.

The most energy-efficient strategy, superinsulation (which reduces a house's energy needs for heating and cooling below a tenth of their usual levels), is still standard only in severe climates, but energy efficiency in general is making progress. Efficiency is taken for granted in new homes, "especially by the better builders," says Larry Sherwood, executive director of the New England Solar Energy Association.

In existing homes, the situation will continue to improve, but more slowly than it has. While "people are not going to rip the insulation out of their walls," as efficiency lobbyist Geller says, they may well loosen their energy belts a notch or two.

According to a study by Battelle Memorial Institute, curtailment of energy services (turning down the thermostat, for example) and lifestyle changes (taking a bicycle instead of a car) have played a greater role in home savings than in any other sector. If home energy bills were to drop dramatically, these energy-saving strategies could easily be abandoned. But the effect of the oil price drop on most energy bills is likely to be minimal, because oil accounts for only about a tenth of home and office energy consumption.

In other areas, progress seems to be more permanent. "Where industry has actually invested money or adopted house-keeping procedures that promote energy conservation, I doubt they will be dropped," says the Sierra Club's Yeager. He suggests that the more expensive strategies will no longer seem as attractive, however, and that the rate at which new energy-saving strategies are adopted will decline. Geller is more optimistic: "There's a lot of savings with two-year paybacks that haven't been made yet," he says, but he concurs that the rate of improvement will drop.

Electricity is in a very different economic position from other forms of energy. Electricity prices, three quarters of which are based on fixed costs such as capital, labor, and profit margin, have kept slightly ahead of inflation in the past four years. As new capital-intensive plants come on line, those fixed costs will continue to increase, raising rates and creating further incentives for customers to use electricity more wisely. "The drop in oil prices may shift attention from saving heat to saving electricity," says Lovins. "If so, that's long overdue—electricity is the most costly form of energy."

Such savings would be somewhat ironic for utilities, most of which have a large surplus of generating capacity. During the past ten years, while they were bogged down with huge, over-budget construction projects, many utilities encouraged their customers to use less electricity, or to use it at times when generating capacity was not overtaxed. Now that these new plants are coming on line and boosting rates, customers are scurrying for efficiency improvements—better motor controls, lights, and the like. By doing so they have cut growth in demand to 1.7 percent per year, a quarter of what it was in the 1960s. Customers can save a kilowatt hour for a fraction of the cost of generating one, Lovins says.

What's more, the effects of the 1978 Public Utility Regulatory Policies Act are beginning to be felt. The law requires utilities to buy power generated by wind, small hydroelectric dams, co-generation, and similar privately owned technologies at a fair price. In California, utilities already have firm offers for private power generation equal to about half their peak load. In Maine the figure is 22 percent; in Vermont, 14 percent. And in Texas, private supply will exceed demand for electricity during some hours of the day, says Clark Gellings, head of the Electric Power Research Institute's demand and conservation department.

The Southeast and Northwest have the largest surpluses of electricity, while the Midwest and Northeast are not as well endowed, according to Gellings. But there are enough high-voltage transmission lines to allow for an exchange of much of the surplus. With many utilities offering their excess power at bargain rates, utilities see fewer incentives to promote conservation.

Utilities should not abandon conservation, however. The cost of electricity-saving devices is less per kilowatt hour saved than the cost of producing a kilowatt hour of electricity—even if the plant cost nothing to build, Lovins says. By his calculations it would be cheaper for them to pay for energy-saving devices in homes, factories, and offices than to pay for fuel, operations, and maintenance at their plants.

Besides, efficiency programs reduce the uncertainty utility planners face, says Ralph Cavanagh, a senior attorney with the Natural Resources Defense Council. People will continue to use energy more efficiently; by helping them to do so, a utility can better predict demand, he says.

In the long term, the key to making conservation programs useful is to incorporate them into utility planning—to set and meet goals for conservation and thus eliminate the need to build new power plants when the surplus disappears. Some power companies taking this rigorous approach to conservation are Pacific Gas & Electric, Southern California Edison, Bonneville Power Administration, the Tennessee Valley Authority, Florida Power & Light, and General Public Utilities, the owners of Three Mile Island. A 1983 survey by the Electric Power Research Institute found that 300 utilities were operating more than 1,000 efficient-use and load-management programs, but most of them still do not take efficiency seriously enough. Some have started to encourage use of electricity during times of the day when their generators are idle.

Bonneville expects to conserve enough energy to displace four large power plants by 2004. Meanwhile, it has a power surplus that is expected to last into the 1990s, which it plans to sell to California while putting conservation programs on a back burner for a few years. Lovins thinks Bonneville should proceed with its conservation efforts, however, concentrating on commercial and industrial savings, which cost half a cent per kilowatt hour, instead of residential savings at two or three cents. Lovins says the half cent per kilowatt hour is less than it costs the utility to run its steam power plants.

Among the utilities' biggest challenges are spreading the word about conservation and making sure people can afford the improvements. Pacific Power & Light is conducting an experiment in Hood River County, Ore. The utility has outfitted some 3,000 homes with as much weatherization as was deemed cost-effective, including fiberglass in the ceilings and basements and triple-glazing on the windows. This cost the utility an average of $4,300 per home, which will likely translate into two or three cents for each kilowatt hour it saves—less than it would have cost to generate that power. Utility officials may replicate the experiment elsewhere if it proves successful.

Florida Power & Light has also built energy savings into its resource plans. Some successful programs include rebates of up to $600 for replacing an old, inefficient air conditioner—sweetened with a $75 rebate to the contractor who sells the replacement. Similar incentives are provided for attic insulation, window film, and solar water heating.

Such efforts are still the exception. The country has a long way to go before it can boast of a truly energy-efficient economy. But falling oil prices are not likely to keep nationwide efficiency levels from growing. After all, the oil glut is a sign of energy conservation's success, not a harbinger of its demise.

The shift to an energy-efficient economy is well under way in the United States. From light bulbs to automobiles, from refrigerators to furnaces, energy-efficient devices have become a part of the U.S. economy. Even though some factors—such as the oil crash and the utilities' surpluses of generating capacity—are hindering the spread of energy efficiency, many mechanisms already in place will carry forward the advances of the last several years.

Changes in patterns of energy use affect deeply ingrained ways of living and working, so it's not surprising that they take a long time to happen. The transition from wood to coal took about 50 years, as did the shift from coal to oil and natural gas. It would be unwarranted to lose patience after only 13 years on the path to sustainable energy use.

ENERGY EFFICIENCY: LESS MEANS MORE[6]

How the world fuels its growth is central to many of earth's most critical problems, especially its environmental health. Increasing evidence links fossil fuel emissions to acidifying lakes, dying forests, reduced crop yields and human respiratory disease. "Our energy systems are irrevocably altering the climate by adding 5.4 billion tons of carbon to the atmosphere each year, more than a ton for each person on the planet," notes Christopher Flavin, an energy analyst with the Washington, D.C.-based Worldwatch Institute. "Simply put, an environmentally sound energy strategy is a prerequisite to a sustainable society." And a prerequisite to any environmentally sound energy strategy, he and others argue, must be more efficient use of energy.

Since the Arab oil embargo nearly 15 years ago, a quiet but dramatic revolution in energy conservation has swept the industrialized world, reducing the projected increase in atmospheric pollutants and in large part producing the world oil glut. "The world has saved far more energy [since 1973] through improved efficiency than it has gained from all new sources," Flavin and Alan B. Durning write in a new Worldwatch Institute report, "Building on Success—the Age of Energy Efficiency."

Yet as experts survey prospects for the 21st century, they see an even more dramatic need for saving energy. Energy waste threatens the economic health of societies large and small, industrial and agrarian, socially planned and market driven. Although new energy-conservation technologies loom on the horizon, institutional resistance, inadequate consumer education and old ways of doing business threaten to slow their adoption.

The United States currently spends about 11.2 percent of its gross national product on energy, while Japan spends only about 5 percent. This relative inefficiency in fueling its energy needs costs the United States $220 billion a year, according to Arthur H. Rosenfeld, director of the Center for Building Science at Lawrence Berkeley Laboratory (LBL) in Berkeley, Calif. Moreover, Rosenfeld points out, this cost differential gives the Japanese

[6]Reprint of a May 7, 1988 article by Janet Raloff, *Science News* staffwriter. Reprinted by permission from *Science News*, the weekly magazine of science. Copyright © 1988 by Science Service, Inc.

"about a 5 percent economic edge on everything they sell"—both in the United States and in foreign competition with U.S-made products.

This is not to downplay recent progress made in the United States in reducing energy use, says Rosenfeld, whose center is renowned for pioneering energy-conserving technologies. It just illustrates that there is room for much improvement.

How much? Energy analyst John O. Blackburn, Duke University's Distinguished Service Professor Emeritus of Economics, says he believes inefficiencies may still account for 50 percent of the energy used in the United States and most Western European countries, somewhat less than 50 percent in Japan, and considerably more than 50 percent in the Soviet Union.

Flavin and Durning offer incentives and guidelines for tackling these inefficiencies in their report. They point out that the energy saved by Western industrialized countries and Japan exceeds the energy used by Africa, Latin America and South Asia combined, and represents the single largest step toward reducing their dependence on oil imports. In fact, Durning says, "we feel it's now possible in most industrialized countries to keep energy consumption level for the foreseeable future"—without sacrificing economic growth.

Just a decade ago, many energy analysts predicted that efficiency improvements of 20 or 30 percent could be achieved only at the expense of wrenching societal change—such as shivering through winters in the dark or abandoning the family automobile. Now, Flavin and Durning report, these improvements have occurred, but largely without notice. Between 1973 and 1985, most Western European nations reduced energy consumption—as measured per dollar of gross national product—by 18 to 20 percent. The United States and Japan saved even more, 23 and 31 percent respectively.

Strategies

One way to discourage waste is to raise energy prices with user taxes—a policy many nations apply to gasoline. Taxes on U.S. gasoline average about 29 cents per gallon—very low by Western European standards. However, U.S. gasoline is taxed only to generate revenues for road construction and maintenance. Durning points out that Denmark directs its $2.93 tax on each gallon at discouraging gas use. And he notes that because Sweden's gasoline tax is inversely related to the price of oil, it provides a steady conservation incentive. As the price of oil declines, the Swedish tax goes up.

Efficiency standards are an alternative where a tax can't be applied, would not motivate effectively, or would hit certain segments of society unfairly. U.S. major appliance rules, due to go into effect next year, set minimum efficiency limits for new refrigerators, freezers, air conditioners, heat pumps, water heaters, gas ranges, dish washers, clothes washers, furnaces and boilers.

These standards should trim electricity consumption 30.4 terawatt-hours (trillion watt-hours) by 1995 and reduce peak demand by 12,800 megawatts, according to calculations by Howard Geller, associate director of the Washington, D.C.-based American Council for an Energy Efficient Economy (ACEEE). Electric water heaters, gas water heaters and refrigerator/freezers are expected to account for the biggest savings—about 20 percent each by the year 2000.

Most electric utilities are adopting time-of-day and peak-load pricing as conservation incentives. Large, "base load" coal and nuclear plants tend to produce the cheapest electricity. Because producing small increments above that base load costs more per kilowatt, users who contribute to demand peaks—consuming power around midday or throughout the hottest days in air conditioning season—are often charged more.

Newer to utility markets is least-cost planning. Traditionally, when projections indicated that demand for electricity would outstrip a utility's capacity to supply it, the utility ordered new generating equipment. Today, many states require utilities first to prove that new equipment will cost less than other options—such as investing in efficiency, initiating time-of-day pricing, buying power from independent generators (including wind-power-plant owners) and cogenerating heat and electricity.

Since January 1986, 37 states have taken 61 different actions to explore or promote least-cost electrical planning, according to Nancy Hirsh of the Washington, D.C.-based Energy Conservation Coalition (ECC). A December ECC study found that at least 14 states and the District of Columbia directly encourage utility investments in energy-efficiency programs through least-cost measures.

Technologies High and Low

Among other innovative concepts for increasing energy efficiency are:

• *Planting trees:* Stately trees not only shade underlying buildings, but also provide regional evaporative cooling as they transpire. The cooling effect is especially useful in limiting the development of urban "heat islands"—inner-city regions whose asphalt and dark-roofed surfaces serve as unintended heat stores. Heat islands can dramatically increase air-conditioning use. But research by Hashem Akbari, Rosenfeld and colleagues at LBL indicates that planting just three trees near a light-colored home can, when done throughout a community, reduce residential cooling demand on a hot summer day—by 18 percent in Phoenix, Ariz., by 34 percent in Sacramento, Calif., and by 44 percent in Los Angeles.

Shade accounts directly for about a third of the cooling. Evapotranspiration and changing home reflectivity with light-colored roofs and paints play a much greater role. "If you plant three trees about every house in [California's] San Fernando Valley, and whitewash those houses, you can save 2,000 megawatts—or two standard power plants' worth of air conditioning," Rosenfeld told *Science News.* By providing some thermal shelter in winter, these plantings can offer smaller, but still significant, reductions in heating needs.

A less obvious benefit from cooling with trees (instead of with combustion-generated electricity) is reduced production of carbon dioxide—a "greenhouse" gas that contributes to global warming. In a new draft paper, Rosenfeld, Philip Martien and Leo Rainer of LBL conclude that nationwide residential and commercial tree planting could save 0.9 quads (10^{14} btu) of utility-generated energy and thereby reduce atmospheric carbon emissions by up to 18 million tons annually. While others have suggested using forests to trap carbon emissions (SN: 4/30/88, p. 285), Rosenfeld says that because of the fossil-fuel use they save, "our urban trees are about 10 times as effective as forest trees" in limiting atmospheric carbon dioxide buildup.

• *Superinsulating buildings:* Doubling the normal insulation recommended for buildings and including an airtight liner within the wall increases initial building costs by only 5 percent and pays back those added costs in energy savings within five years, Flavin and Durning report. "Energy for a Sustainable World," a September 1987 report by the Washington, D.C.-based World Resources Institute, describes one commercial version. Built near New York City with 120 square meters of floor space, this home could be heated for about 1,400 kilowatt-hours (kwh) of electricity, according to that report—the annual energy consumption of most U.S. refrigerators.

• *Replacing aging major appliances with more efficient ones:* U.S. refrigerators consume about 7 percent of the nation's electricity, the energy equivalent of about 25 large power plants. New units use considerably less. According to ACEEE's John H. Morrill, a 1972 refrigerator with a top-mounted freezer and automatic defrost typically used, when new, about 2,000 kwh of electricity per year. A similar model sold in 1985 runs on about 1,100 kwh a year. And 27 similar models marketed under nine brand names last year exceeded the best efficiency available only a year earlier. Says Morrill, "If all the households in the U.S. had the most efficient refrigerators currently available, the electricity savings would eliminate the need for about 12 large nuclear power plants." Similar savings are possible with other appliances.

• *Using more efficient lighting:* Currently, 20 to 25 percent of U.S. electricity—about 100 power plants' worth—provides illumination. Fluorescent fixtures using the most efficient ballasts (spark discharge devices) consume 25 to 30 percent less electricity than standard fluorescents, and roughly 75 percent less energy than typical incandescents with the same light output.

Testifying in March before the House Subcommittee on Energy and Power, ACEEE's Geller estimated that enacting pending legislation for U.S. ballast-efficiency standards would save 500 billion kwh of electricity between 1990 and 2010—about half of which would have been generated from oil and gas. The oil and gas savings alone would be equivalent to 560 million barrels of oil, meaning "the ballast standards can be viewed as a very large oilfield," says Geller.

And fluorescent-efficiency increases of another 40 percent or more are on the horizon, says Rudy Verderber at LBL's National Lighting Laboratory. For example, when the mercury plasma discharges in a fluorescent light, it generates invisible ultraviolet radiation. Each ultraviolet photon reaching the tube's phosphor-coated inner surface will cause that coating to emit a photon of visible light. However, using a different phosphor, "it's [theoretically] possible to provide two photons of visible light from each photon of ultraviolet," Verderber says—doubling the lamp's efficiency. Efforts to find such a phosphor should begin soon at his lab.

Already underway are several other programs. One seeks to equalize the ratio of mercury isotopes typically pumped into fluorescent tubes. This will limit photon "entrapment" (absorption) by the overrepresented isotopes, Verderber says, for a 3 to 5 percent increase in efficiency. Far more impressive are the potential gains possible with LBL's new "surface wave" fluorescent lamp. Its higher-frequency ballast—cycling at about 100 million hertz, not the 20,000 hertz typical of today's high-frequency ballasts or 60 hertz of conventional fluorescents—eliminates the need for electrodes. Since these are the primary limit to a fluorescent lamp's life, the new models might last five or more times longer, Verderber says. And because this lamp produces ultraviolet emissions closer to the phosphor coating, far fewer ultraviolet photons are lost to entrapment. The result is a 40 to 50 percent increase in efficiency. Finally, because this design eliminates current limitations on a fluorescent's shape, future models might resemble regular incandescents.

• *Coating windows with infrared reflectors:* Today, a third of the heat in U.S. homes escapes through closed windows. That is because the standard insulated wall has an insulating "R value" of 11, but the typical single pane of glass is a mere R-1. Double-glazed windows, with an R-2 insulating value, can be upgraded to R-4 by coating the inner surface of one pane with an infrared-radiation (heat) reflector, such as tin oxide, and filling the space between the panes with argon gas. Developed at LBL, such windows are commercially available.

But LBL now has experimental "triple-glazed" windows with an insulating value of R-6 to R-10. Its latest models cover the inner surface of each exterior pane with an infrared-reflecting coating and fill the insulating space between each pane with krypton—a gas about 60 percent less conducting than air and twice as insulating as argon. According to LBL's Dariush Arasteh, commercially prepared experimental versions of this window should debut in demonstration homes later this year.

• *Installing power electronics:* Today, most electrical equipment is configured to be either fully on or fully off. Yet most demands on them need only some fraction of full capacity. For example, when a commercial process needs to reduce the flow of a liquid through a pipe, most systems use valves to restrict flow while the pump's motor runs on at full speed. Varying pump speed might save 25 to 40 percent of the energy and accomplish the same thing.

Between 50 and 60 percent of U.S. electricity currently drives electric motors, most of which run at fixed speeds and voltages. Power-electronics devices—from light dimmers to variable-speed drives—can better match electrical devices and appliances to their varying needs, and save energy in the process.

Unfortunately, designers of most offices, industrial processes and residential appliances have failed to keep up with recent advances that have dramatically cut their size, cost, weight, inefficiencies and unreliability, says Ralph J. Ferraro, manager of power electronics control systems for the Palo Alto, Calif.-based Electric Power Research Institute (EPRI). To identify where new generations of such devices might play a role, EPRI founded the Power Electronics Applications Center (PEAC) in Knoxville, Tenn. Now an independent applied-research center, PEAC works with EPRI, universities, national laboratories, manufacturers and trade associations to spur greater use of these energy-saving and productivity-enhancing devices. Only about 10 to 15 percent of U.S. electricity now flows through power-electronics systems, but "it's anticipated that by the turn of the century, some 50 to 60 percent of all U.S. electricity will," Ferraro says.

Reducing Obstacles

These and other technologies indicate room for further energy savings. But potentially formidable social obstacles are holding back adoption of many improvements. For example, while high-efficiency water heaters may cost only slightly more than inefficient ones, many consumers won't buy them because they lack the information needed to calculate how quickly energy savings will pay back the extra cost. Similarly, many of the least-energy-efficient industries (like housing) are fragmented. This makes it hard for them to institutionalize change and to conduct the long research programs that pay off in more energy-efficient materials. Finally, many governments lack the economic and environmental data that might serve as incentives for imposing measures to encourage energy conservation.

Several organizations are working to counter these obstacles. ACEEE, for example, publishes an annual brand-name and model-number list of the most efficient major residential appliances in the United States. It also provides instructions for computing how quickly a more efficient unit will pay back its higher cost.

Researchers and policy analysts have attempted to persuade Congress that increasing federally funded energy-conservation research budgets will pay handsome dividends—especially in areas like housing. For example, Rosenfeld testified last year that federal funding for U.S. energy-conservation projects is declining—from $344 million in 1980 to $162 million in 1987—despite the many examples of projects whose technologies, once widely adopted, "will yield savings of 1,000-to-l." Initial implementation of some new technologies by the building sector is already yielding energy savings of several billion dollars a year, Rosenfeld says. Within a few decades, he predicts, they will save U.S. energy consumers roughly $100 billion annually.

And finally, a spate of new studies have begun quantifying environmental and economic repercussions of measures affecting energy use. For example, "Money to Burn?," an October 1987 World Resources Institute study by Mark Kosmo, analyzed costs associated with energy subsidies. His data, reflecting measures in 30 countries, forcefully support the axiom that where energy costs more, consumers tend to use it more efficiently. Similarly, "Acid Rain and Electricity Conservation," published last year by ACEEE, says that both the environment and the pocketbook would benefit from saving energy. The report's data came from an analysis of East Central states. Geller, one of the study's authors, believes its findings offer an encouraging message: "Use of emission ceilings, along with the least-cost approach to acid-rain control, could cut energy use for electricity production by 10 to 20 percent, and save states and their consumers tens of billions of dollars."

Although improving energy efficiency lacks the glamour of fusion and space-based solar collectors, Flavin and Durning say, "perhaps no other endeavor is as vital to the goal of fostering sustainable societies." Moreover, the Worldwatch analysts argue, because most energy use carries a high environmental cost, "without improved efficiency, it is only a question of which will collapse first: the global economy or its ecological support systems. With greater energy efficiency, we stand at least a fighting chance" of staving off the collapse of both.

BIBLIOGRAPHY

An asterisk (*) preceding a reference indicates an excerpt from the work has been reprinted in this compilation.

BOOKS AND PAMPHLETS

Adamson, Greg. We all live on Three Mile Island: the case against nuclear power. Pathfinder Press. '81.

Addinall, Eric and Ellington, Henry. Nuclear power in perspective. Nichols. '82.

Alm, Alvin and Curham, Joan. Coal myths and environmental realities. Westview. '84.

Arnold, Guy. Coal. Gloucester. '85.

Aviel, S. Daniel. The politics of nuclear energy. University Press of America. '82.

Bacon, Hilary and Valentine, John. Power corrupts: the arguments against nuclear power. Pluto. '81.

Ballard, Steven S. Water and western energy: impacts, issues, and choices. Westview. '82.

Cable, Charles R. Nuclear energy. Raintree. '83.

Catherall, Ed. Solar power. Wayland. '81.

Colglazier, E. W. The politics of nuclear waste. Pergamon. '82.

Cook, Joseph L. and Cook, Earleen H. Coal gasification as an energy alternative. Vance Bibliographies. '80.

Coombs, Charles. Coal in the energy crisis. Morrow. '80.

Croall, Stephen. Nuclear power for beginners. Pantheon. '83.

Daniels, Farrington. Direct use of the sun's energy. Yale University Press. '83.

Davis, Bertha and Whitfield, Susan. The coal question. F. Watts. '82.

Dineen, Jacqueline. Energy from the sun, wind, and tide. Endlow. '88.

Dineen, Jacqueline. Nuclear power. Enslow. '88.

Dolan, Thomas J. Fusion research: principles, experiments and technology. Pergamon. '82.

Evans, Nigel and Hope, Chris. Nuclear power: futures, costs, and benefits. Cambridge University Press. '84.

Farhar-Pilgrim, Barbara and Unseld, Charles T. America's solar potential. Praeger. '82.

Fenn, Scott. The nuclear power debate. Praeger. '81.

Fisk, Marian and Anderson, H. C. Introduction to solar technology. Addison-Wesley. '82.

Flavin, Christopher. Reassessing nuclear power. Worldwatch Institute. '87.

Flood, Michael. Solar prospects. Wildwood. '83.

Fogel, Barbara R. Energy choices for the future. F. Watts. '85.

Gard, H. P. Advances in solar energy technology. Reidel. '87.

Garrison, Jim. The plutonium culture: from Hiroshima to Harrisburg. Continuum. '81.

Goldschmidt, Bertrand. The atomic complex: a worldwide political history of nuclear energy. American Nuclear Society. '82.

Goode, Stephen. The nuclear energy controversy. Watts. '80.

Grenon, Michel. The nuclear apple and the solar orange. Pergamon. '81.

Gross, Robert A. Fusion energy. '84.

Grossman, Karl. Cover up: what you are not supposed to know about nuclear power. Permanent Press. '82.

Haines, Gail. The great nuclear power debate. Dodd, Mead. '85.

Hawkes, Nigel. Nuclear power. Gloucester. '84.

Hawkes, Nigel. Oil. Gloucester. '85.

Hulstrom, Roland. Solar resources. MIT Press. '89.

Hunt, V. Daniel. Handbook of conservation and solar energy. Van Nostrand Reinhold. '82.

Ide, Arthur F. Coal, yesterday's energy today. Ide House. '81.

Ikenberry, G. J. Reasons of state: oil politics and the capacities of American government. Cornell University Press. '88.

Kaku, Michio and Trainer, Jennifer, eds. Nuclear power: both sides—the best arguments for and against the most controversial technology. Norton. '82.

Kaplan, Sheila. Solar energy. Raintree. '83.

Katzman, Martin T. Solar and wind energy. Rowman & Allanheld. '84.

Kiefer, Irene. Nuclear energy at the crossroads. Atheneum. '82.

Kohl, Wilfrid L. After the second oil crisis: energy policies in Europe, America, and Japan. Lexington. '82.

Koral, Richard L., ed. Foundations of the solar future. Fairmont. '81.

Lieber, Robert J. The oil debate. Praeger. '83.

Lyttle, Richard B. Shale oil & tar sands: the promises and pitfalls. Watts. '82.

Mansfield, Jerry W. The nuclear power debate: a guide to the literature. Garland. '84.

McAninch, Sandra. Sunpower: a bibliography of United States government documents on solar energy. Greenwood. '81.

McDaniels, David K. The sun, our future energy source. Wiley. '84.

McKie, Robin. Solar power. Gloucester. '85.

McVeigh, J. C. Sun power: an introduction to the applications of solar energy. Pergamon. '83.

Meltzer, Michael. Passive and active solar heating technology. Prentice-Hall. '85.

Morone, Joseph and Woodhouse, Edward. The demise of nuclear energy? Yale University Press. '89.

Mossavar-Rahmani, Bijan. Lower oil prices: mapping the impact. Energy and Environmental Policy Center. Harvard University. '88.

Mueller, Kimberly J. The nuclear power issue. California Institute of Public Affairs. '81.

Niu, Keishiro. Nuclear fusion. Cambridge University Press. '88.

Nussbaum, Bruce. The world after oil. Simon and Schuster. '83.

Odell, Peter R. and Rosing, Kenneth E. The future of oil. Nichols. '83.

Pasqualetti, Martin and Pijawka, K. D. Nuclear power: assessing and managing hazardous technology. Westview. '84.

Payne, Sherry. Wind and water energy. Raintree. '83.

Penner, S. S. New sources of oil and gas. Pergamon. '82.

Pleatsikes, Christopher, et al. Solar energy and the U. S. economy. Westview. '82.

Pringle, Laurence. Nuclear energy: troubled past, uncertain future. Macmillan. '89.

Rice, Dale. Energy from fossil fuels. Raintree. '83.

Rippon, Simon. Nuclear energy. Heinemann. '84.

Schobert, Harold H. Coal, the energy source of the past and future. American Chemistry Society. '87.

Scott, Elaine. Oil: getting it, shipping it, selling it. F. Warne. '84.

Simeons, Charles. Hydro-power: the use of water as an alternative source of energy. Pergamon. '80.

Skousen, Eric N. The war against nuclear power. Freeman Institute. '82.

Smallwood, Kenneth. Water: a source of energy today and tomorrow. Ide House. '82.

Wagner, Henry and Ketchum, Linda. Living with radiation: the risk, the promise. Johns Hopkins University Press. '89.

Walton, A. L. and Warren, E. H. The solar alternative: an economic perspective. Prentice-Hall. '82.

Weiss, Ann E. The nuclear question. Harcourt Brace Jovanovich. '81.

Yates, Madeline. Sun power: the story of solar energy. Abingdon. '82.

For those who wish to read more widely on the subject of energy and conservation, this section contains abstracts of additional articles that bear on the topic. Readers who require a comprehensive list of materials are advised to consult the *Readers' Guide to Periodical Literature* and other Wilson indexes.

OIL

Foreign oil: the scourge of the '70s may be a blessing now. Barbara Starr and Tim Smart *Business Week* 40 Mr 23 '87

Domestic oil producers and government officials have expressed alarm about America's growing dependence on foreign oil, but their fears may be exaggerated. Despite its professed concern about the effects of energy dependence, the Reagan administration believes that as long as the United States isn't vulnerable to oil blackmail, a 50 percent dependence on foreign oil poses no danger. Proponents of this argument point to OPEC's decreased power and to major oil imports from Canada and Mexico. The U. S. economy is better off using cheap imported oil as long as U. S. production costs remain high. Fearful of domestic energy programs that would cost vast sums of money, the administration will probably do little to stimulate the ailing oil industry beyond offering shopworn proposals like decontrolling natural gas prices, opening more federal lands to exploration, and easing environmental restrictions.

A savage Rocky Mountain low that won't let up. William C. Symonds *Business Week* 112–13+ O 31 '88

A severe economic recession still grips the Rocky Mountain states of Montana, Idaho, Wyoming, Utah, Colorado, and New Mexico, and there is little hope that the situation will improve in the near future. The region once had a thriving economy driven by the oil and mining industries, but since the boom collapsed in 1983, the states have lagged behind the rest of the nation in nearly every area of economic growth. Many of the states' residents are fleeing to more prosperous regions. The loss of well-paid energy workers has adversely affected the region's real estate market, causing sharp declines in residential and commercial construction. To promote growth, the Rocky Mountain states are competing against one another to attract businesses and government contracts.

The third oil crisis. James Cook *Forbes* 140:37+ D 14 '87

Given the political volatility of the Middle East, another oil crisis is not beyond the bounds of possibility. The Middle East still produces almost 30 percent of the free world's oil. Most investors and business people take comfort from the U. S. Strategic Petroleum Reserve and are optimistic about the country's ability to weather another crisis. The reserve holds enough oil to produce the equivalent of half the import needs of the United States for four months. It may prove to be an imaginary defense, however. At the end of last year, world oil reserves were no higher than in 1979, and there are considerable doubts about the United States' ability to mobilize its supply. In addition, the stock market crash demonstrated how quickly prices can change. In an emergency, panic buying would create a seller's market, and the government would soon lose control of the situation.

Impending United States energy crisis. Robert L. Hirsch *Science* 235:1467-73 Mr 20 '87

The U. S. oil and gas industry has been dramatically weakened by the recent oil price collapse. Domestic drilling activity reached a new post-World War II low during the summer of 1986. Given a weak, unstable oil price outlook, U. S. capability will continue to deteriorate. In the last year U. S. imports of foreign oil have risen significantly, and if market forces alone dominate, U. S. dependence is expected to rise from 32 percent in 1983 to the 50 to 70 percent level in the not-too-distant future. The 1973 oil embargo and the subsequent attempts to improve U. S. energy security vividly demonstrated the huge costs and long periods of time required to change our energy system. These facts, coupled with the nation's generally short-term orientation, suggest a strong likelihood of a new U. S. energy crisis in the early to middle 1990s.

Low-cost oil is fueling energy worries (report by the Department of Energy). *Science News* 131:201 Mr 28 '87

The Department of Energy's Energy Security report predicts that Persian Gulf countries may supply 30 to 45 percent of the world's oil within eight years. Petroleum accounts for 40 percent of U. S. energy. According to the report, declining oil prices have encouraged greater oil use, caused reductions in oil exploration, and forced some U. S. production operations out of business. In the March 20 issue of *Science*, Robert L. Hirsch, vice president of research for ARCO Oil and Gas, suggests ways to reduce U. S. dependence on foreign oil, including an oil import tax, an oil import quota, tax incentives for U. S. exploration and production, and incentives for encouraging a shift to other energy sources.

The two George Bushes (views on oil prices). Harold Evans *U. S. News & World Report* 100:76 Ap 21 '86

Vice President George Bush has taken a Jekyll-and-Hyde approach to declining oil prices. The former defender of free enterprise and market forces went to Saudi Arabia to tell King Fahd to cut oil production so that prices would stop falling. After the meeting, Bush stated that the United States wants the lowest possible oil price as long as it favors the domestic oil industry and national security. Bush is wishing for relief through manipulation of the market. There are economic and political benefits to the reduced oil prices, and it is against U. S. interests to try to alter the market. Instead of taking a global approach to the problem, the United States should work to develop specific solutions for troubled areas. Increasing the federal gasoline tax, promoting the use of nonoil energy sources, and rescheduling U. S. oil producers' bank loans would all discourage import dependence while preserving the benefits of low-priced oil.

Conservation up front. Gaylord Nelson *Wilderness* 51:2 Wint '87

The best response to the threat of another energy crisis is a renewed commitment to alternative energy sources and conservation, but the Reagan administration has focused instead on the exploration and exploitation of oil and gas reserves. Interior Secretary Donald Hodel has targeted the Arctic National Wildlife Refuge in Alaska, an area set aside by Congress as wildlife habitat, for oil exploration; yet even if commercially producible oil were found there it would provide only a negligible portion of the country's oil needs. Before decisions are made that could irreversibly change our public lands, a long-term energy policy must be established that recognizes the importance of energy conservation alternatives.

Nuclear Energy

Nuclear power. David J. Rose *The Bulletin of the Atomic Scientists* 41:76–8 Ag '85

PART OF A SPECIAL FORTIETH ANNIVERSARY ISSUE ON THE FIRST USE OF THE ATOMIC BOMB. Although there are many operating nuclear reactors in the United States, projections for 1000 gigawatts (GW) of power from nuclear sources by the year 2000 have been trimmed to 125 GW or less. Problems with technology, communication, and public opinion have hampered nuclear power development in the United States more than in other countries. The growing reliance on electrical power is favorable for developing wind, geothermal, and photovoltaic power, as well as nuclear sources. Permanent entombment of used fuel, rather than reprocessing, would break the connection between nuclear power and nuclear weapons. Small, fuel efficient, light-water and high-temperature gas reactors being developed in Germany and Japan represent the next generation of nuclear plants.

Chernobyl's other legacy: shifting public perceptions of nuclear risk (with editorial comment by Gilbert Fowler White). Christoph Hohenemser and Ortwin Renn *Environment* 30: inside cover, 4–11+ Ap '88

Studies of the 1986 Chernobyl nuclear accident have tracked its effects on world health, public opinion of nuclear policy, and nuclear risk-management strategies. The accident, which involved the largest release of radioactivity from a single technological disaster, produced a varied pattern of fallout, with major hotspots directly related to local rainfall. Immediate fatalities numbered 31, but some predict as many as 28,000 related cancer fatalities worldwide during the next 50 years. Public opposition to nuclear power increased after the accident in proportion to each country's level of fallout exposure. In subsequent months, this new opposition decreased or plateaued in relation to the maturity of nuclear debate in individual countries. The transnational nature of the fallout caught many nations off guard, forcing them to improvise emergency strategies. This suggests that further studies of risk-management strategies are needed.

Nuclear bankruptcy. *Fortune* 117:14 Ja 18 '88

Nuclear power plants figure prominently in the troubles of four power companies that could be headed for bankruptcy. Public Service of New Hampshire, owner of the Seabrook plant, and Long Island Lighting, owner of the Shoreham reactor, have been blocked from bringing their plants on-line by state regulators who have environmental and economic concerns. In the Mississippi Delta, state regulators have denied Gulf State Utilities and Middle South Utilities the rate increases that would have permitted them to recover the costs of their nuclear power plants. Utilities that have nuclear power plants operating have other worries: the prospect of plant shutdowns when licenses expire and the attendant cost of dismantling reactors.

Not so bad after all? Mark Miller *Newsweek* 112:65 Jl 25 '88

Rising concern about the greenhouse effect has brought hope to the embattled nuclear power industry. Some longtime critics of nuclear energy, which unlike fossil fuels does not produce carbon dioxide or other greenhouse gases, concede that nuclear power deserves another look. They caution against rushing to embrace a technology that has potentially disastrous consequences of its own, however. Most experts agree that future plants will have to be smaller to be competitive, with custom-designed plants giving way to a few standardized designs that can be constructed quickly and cheaply. There is also the issue of radioactive waste disposal to consider. In addition, some environmentalists are worried that a renewed nuclear program could siphon off funds needed to develop renewable resources such as solar, geothermal, and wind energy.

Nuclear reactions. Joel Davis *Omni* 10:40-2+ My '88

Rumors of the death of the nuclear power industry appear to be premature. No nuclear power plant has been ordered in the United States since 1978, but overseas demand for American reactors is increasing. In 41 countries, a total of 590 plants are in development, 70 percent of which are modeled on the standard U. S. light-water reactor (LWR). Meanwhile, U. S. demand for power has increased about 40 percent since 1973. Since the Three Mile Island disaster, which involved an LWR, some professors and researchers have questioned whether America should again accept LWRs. They place their hopes in so-called inherently safe reactors, which come in three types: integral fast reactors, high-temperature gas-cooled reactors, and process inherent ultimate safe reactors. Even more sophisticated designs are under development, but critics charge that no reactor can be made inherently safe.

Shut down but not out (Seabrook and Shoreham plants). Harvey Wasserman *The Progressive* 52:20-1 Ag '88

Despite two recent victories for opponents of nuclear energy, the threat of a nuclear accident remains high. Recently, the openings of New York's Shoreham plant and New Hampshire's Seabrook plant were successfully blocked by antinuclear activists and by laws that require suitable evacuation plans for the sites.

It appears unlikely that the two plants will ever operate. Their fates mark a major victory for the grass-roots antinuclear movement. All signs indicate, however, that the world's 350 operational commercial nuclear reactors are growing more dangerous. There is a 50 percent chance that a Chernobyl-size accident will occur at an American commercial reactor within the next 20 years, according to James K. Asselstine, a former member of the U. S. Nuclear Regulatory Commission.

Nuclear reactor safety assailed in report (GAO report). Rick Weiss *Science News* 132:213 O '87

According to a report by the General Accounting Office, U. S. nuclear power plants have operated for years under inadequate safety guidelines that have been applied inconsistently. The report, which was requested by Sen. Alfonse M. D'Amato of New York, states that officials of the Nuclear Regulatory Commission (NRC) lack standards for shutting down plants that have been deemed unsafe. While some are shut down, others are allowed to operate despite having similar safety violations. It also criticizes the NRC's failure to promptly resolve general safety issues, such as design, construction, or operating problems that affect numerous plants. The study recommends that the NRC further centralize its management of the national nuclear industry, which would help consolidate information on safety violations. NRC officials strongly contest the report's findings. D'Amato is citing the report as a reason to keep New York's Shoreham nuclear power plant from coming on line.

We are in a heap of trouble (Seabrook). Joelle Attinger *Time* 130:114 O 26 '87

The Seabrook nuclear power plant, a project first announced in 1972, sits idle on the New Hampshire coastline, a victim of cost overruns and environmental battles. Public Service of New Hampshire, Seabrook's principal owner, has already spent $2.1 billion on the plant and recently defaulted on a $37.5 million interest payment, risking bankruptcy. Originally budgeted at $973 million, Seabrook's cost has now reached $5.5 billion, and it has yet to generate any electricity or revenues. Public Service is fighting with creditors over possible rescue plans. One proposal includes a provision that would raise electric rates by 15 percent; it would give Public Service the highest residential electricity rates of any New England utility.

Nuclear sites need better protection (views of Philip Farley). *USA Today* 113:6 Ap '85

Nuclear power plants are a tempting target for terrorists and enemy foreign governments, for security measures are generally inadequate. Philip Farley, special representative for nonproliferation matters during the Carter administration, notes that a terrorist attack could damage a facility and endanger the local population. Terrorists could conceivably steal nuclear fuel to build nuclear bombs. The proliferation of nuclear technology and fuel to unstable governments such as Libya poses additional danger.

NATURAL GAS, COAL, HYDROELECTRICITY

Big energy supplies from little dams. *The Futurist* 19:50-1 Je '85

Small-scale hydroelectric plants will be an important energy source in the future. Developing countries will be able to build systems quickly and relatively cheaply to keep pace with their increasing energy demands; operation of the facilities will also be inexpensive. China has built over ninety thousand such plants powering hydro turbines and generators and is instructing several nations how to harness their water resources the same way. Watershed management is required to prevent overbuilding of the dams in crucial areas, which could cause environmental damage. Even so, there is much room for the development of safe hydroelectric systems; in 1980, only 22 percent of the world's exploitable water power was used to generate electricity.

World supplies of natural gas. Philip H. Abelson *Science* 228:1263 Je 14 '85

Worldwide, natural gas is replacing petroleum, especially in less-developed countries (LDCs). Petroleum reserves have not increased in the last ten years, while natural gas reserves have doubled. More natural gas will likely be found because of improved geophysical prospecting methods and the willingness of oil companies to search for alternative energy sources. Unlike petroleum, natural gas can form from a variety of organic matter and remain stable over a wide range of temperatures. Reserves have been located in thirty countries where no petroleum has been found. Local resources of natual gas are a cost-effective alternative for LDCs. Recognizing this, the World Bank is assisting twenty countries in the exploration and development of local natural gas reserves.

Technologies for clean use of coal. Philip H. Abelson *Science* 229:819 Ag 30 '85

Clean Coal Use Technologies, a recent report by the Energy Research Advisory Board of the Department of Energy, evaluates the problems caused by present energy sources and describes new technologies. Seventy percent of SO2 and 20 to 25 percent of NOx emitted in the United States result from coal-fired utility boilers. A new and inexpensive technology called Limestone Injection Multistage Burner is effective in reducing both emissions and may have broad applications for industry. Much of the sulfur content of Midwestern coal is bound within the mineral pyrite, which can be largely removed by a number of physical methods. Fluidized bed combustion is the optimal technology for new boilers used for industry and utilities. The DOE's $60 million budget for demonstration plants is insufficient in light of the importance of reducing coal-based pollution.

Methane, methane, where from art thou? (atmospheric methane from fossil sources). *Science News* 133:251 Ap 16 '88

During the last decade, scientists have been concerned about rising atmospheric levels of methane, natural gas's chief component, which is accumulating in the atmosphere and contributing to the greenhouse effect by trapping energy from the earth. In the April 7 issue of *Nature*, researchers from the Institute of Nuclear Science in Lower Hutt, New Zealand, report that about one-third of the methane in the atmosphere comes from so-called fossil sources. Formed millions of years ago from decaying plants and animals, fossil methane is found in pockets of porous rock. The percentage of fossil methane in the atmosphere has grown because of increased mining operations and use of natural gas, according to Ralph Cicerone of the National Center for Atmospheric Research in Boulder, Colorado.

A mild alternative? Tim Beardsley *Scientific American* 257:37+ D '87

The Department of Energy (DOE) has awarded four contracts worth $13.5 million over the next three years for studies of mild gasification, a process that involves heating coal to between 1,100–1,500 F in the absence of oxygen. The technique produces a fuel gas, liquid hydrocarbons, and a solid residue called char. The fuel gas can be used in industry. According to Robert O. Ness, Jr., of the Energy Research Center at the University of North Dakota, the liquid hydrocarbons could be used in specialty chemical production. They might also be used as fuels for jets or locomotives. The char may be used for smelting taconite and producing activated carbon, carbon fibers, graphite electrodes, and hydrogen. If the economic prospects look favorable, the DOE plans to commission a mild gasification demonstrator plant capable of processing 1,000 tons of coal daily.

Shattered hopes for synfuels (Great Plains project). Barbara Rudolph *Time* 126:35 Ag 19 '85

Beulah, North Dakota, was a sleepy country town until 1978, when the Department of Energy announced that it would back a $2.1 billion commercial synthetic-fuels plant on the outskirts of town. Then Beulah became a boom town. Today it faces disaster. The decreasing price of foreign oil makes the price of synthetic fuel uneconomical, especially in the face of a ballooning federal deficit. Last month the House of Representatives eliminated all funding for the Synthetic Fuels Corporation. At the same time, the Department of Energy withdrew $1.4 billion in aid to Beulah's Great Plains Gasification Project. Advocates of synthetic fuels argue that oil prices could rise again and that the development of alternative energy sources is good insurance, but Washington clearly thinks that the insurance is too expensive.

Is coal the fuel of the future? Veronique Maurus *World Press Review* 33:51 O '86

Excerpted from the July 22 edition of *Le Monde* of Paris. Coal is the global energy source of the future. Reserves of coal are dispersed across every continent, with more than a 200-year supply in currently known and recoverable reserves. New methods of low-cost recovery are being explored not only in the four traditional exporting countries—the United States, Australia, South Africa, and Canada—but also in new arrivals like Colombia and China. Coal's bright future has shaken the industry's structure. Today 26 companies control two-thirds of coal exports. World-class oil companies like Shell and British Petroleum are establishing themselves at every level of the coal trade. They are aware that new technologies like liquefaction and gasification may make coal a substitute for oil.

SOLAR ENERGY, NUCLEAR FUSION, CONSERVATION

Solar power: alive, well—and almost making money. Mimi Bluestone *Business Week* 132–3 Jl 18 '88

Solar power is on the threshold of significant profitability in disparate U. S. industries. Among its institutional applications, solar power is used in microwave transmitters, oil-drilling instruments, navigation lights, and military communications equipment. Solar cells are also found in consumer products ranging from calculators and radios to blenders and patio lights. The technology is still too expensive for use by utilities, however. Several industry groups are currently developing technologies to reduce sunlight-conversion costs, but their efforts are hindered by federal cutbacks in funds for photovoltaic research. The U. S. share of the world photovoltaic market dropped from 60 percent in 1982 to 30 percent last year. Japan and West Germany, meanwhile, are spending significant amounts on solar research and development.

Making a sun (views of T. A. Heppenheimer). *The Futurist* 19:51-2 Je '85

Fusion energy will probably be an important source of power in the future, but testing is currently hampered by high costs and low output. In order to create fusion energy, hydrogen nuclei are combined with other atoms, which causes a release of energy. But tremendously high temperatures—up to 100 million degrees—are needed to create the reaction. At Princeton Plasma Physics Lab, where the experiments are being conducted, monthly power bills average about $250,000. A Soviet-built toroidal magnetic chamber, or tokamak, holds the generated plasma, or ionized gas, within a magnetic field. So far, the energy output is very low. But some day, according to T. A. Heppenheimer, author of *The Man-Made Sun*, fusion energy could power nuclear reactors.

End of the profligate age? Christopher Flavin and Alan B. Durning *International Wildlife* 18:12-16 My/Je '88

Many nations have made great strides in energy conservation in the past 15 years, but no country has even begun to tap the full potential for further improvements. Thanks to new technologies now appearing on the market, efficiency gains of at least 50 percent over current levels can be achieved in every sector of the global economy. The industrial sector has led the way in improving energy efficiency, but electromechanical drives, electrolysis, and heating can all be made more efficient. Cogeneration also affords a great opportunity to improve industrial energy efficiency. By investing more in these and other improvements, nations would be investing in reduced oil dependence, less air pollution, and control of the global climate. Governments could promote energy conservation by offering more incentives to invest in efficiency and by making energy policy the centerpiece of national and international economic philosophy.

Climate turns chilly for European fusion program. David Dickson *Science* 241:154 Jl 8 '88

The research ministers of the 12 member states of the European Economic Community (EEC) agreed on June 29 to impose the first significant budgetary constraint on the joint European fusion program since the program's inception in the 1950s. According to officials with the EEC's Commission in Brussels, the change will represent a 4 percent cut from current spending plans over the next four years. Fusion scientists say that the impact will be significant because a large proportion of the budget is already committed to meeting fixed costs. Britain led the effort to reduce the budget, but part of the impetus may also have come from a report written by Colin Sweet of the Center for Energy Studies at the Southbank Polytechnic in London. Sweet charged that the economic prospects for fusion have been exaggerated and that the potential environmental risks it poses have been understated.

A good shot at fusion (Particle Beam Fusion Accelerator). A. K. Finkbeiner *Science Digest* 94:20 Ap '86

Physicists at Sandia National Laboratories have taken a step toward controlling fusion. In December 1985, they fired the most powerful charged-particle beam accelerator in history: the Particle Beam Fusion Accelerator II (PBFA II). Fusion is the nuclear reaction that releases the sun's power; densely packed hydrogen nuclei, which usually repel each other, collide under great pressure, and their fusion produces huge amounts of energy.
Scientists are working on two techniques to control the fusion process: magnetic confinement and inertial confinement. Both, however, require more work. PBFA II is an inertial confinement model in which a stacked ring of thirty-six power modules sends out simultaneous beams of charged particles to compress hydrogen. Although the scientists at Sandia have fired the most powerful beam yet, only time will tell which type of machine will produce fusion energy first.

The road to magnetic fusion? (international cooperation). Dietrick E. Thomsen *Science News* 132:294 N 7 '87

A recent report from the Office of Technology Assessment states that controlled thermonuclear fusion could supply a significant portion of the world's electricity by the middle of the 21st century. The report maintains that progress toward this goal will require the internationalization of the magnetic fusion program that the United States has pursued for the last 40 years.
This conclusion is supported by the U. S. Department of Energy, which recently concluded an agreement with the European Community, Japan, and the Soviet Union on cooperation on a project called the International Thermonuclear Experimental Reactor.

Dark days for solar. Bill D'Alessandro *Sierra* 72:34–7 Jl/Ag '87

The solar industry has been devastated by cheap oil, the elimination of tax credits, cuts in federal research funds, and charges of fraud, high prices, and poor quality. By 1980, the solar industry was booming, thanks largely to federal tax credits that allowed deductions of 40 percent of up to $10,000 in purchases of solar or other renewable-energy equipment. By 1986, however, congressional wrangling over tax reform and widespread reports of unscrupulous financing arrangements and fraudulent claims about equipment performance resulted in the dismantling of federal tax credits, even though government reports show that taxpayers received a good return for their solar investments. Today, the solar industry has a proven technology but a tiny market and insufficient funds to promote its products. Without government backing, it will take another oil embargo or shortage to renew the industry.

Solar cells: still a tough sell. Don Best *Sierra* 73:27–9 My/Je '88

The photovoltaic industry has struggled to move ahead, even though it has been abandoned by the Reagan administration and stung by the drop in oil prices. When the United States was in the grip of the early 1970s energy crisis, photovoltaic systems, which use semiconductor materials to convert sunlight into electricity, were touted as the technology that would provide the nation with a clean, reliable, and cheap source of energy. Unfortunately, photovoltaic technology did not advance as rapidly as was anticipated. The lack of government funding for conservation and renewable energy programs since the Carter years has hindered photovoltaic research. Solar-cell production in the United States fell for the first time in 1985 and 1986. Despite the slump, the industry has forged ahead, bolstered by technological breakthroughs and by lower prices, which have opened up new markets for solar energy.

Gone with the wind (tax incentives expire for alternative energy producers). Jacob V. Lamar, Jr. *Time* 127:23 Ja 20 '86

With the expiration of federal tax incentives on December 31, the alternative energy industry and the homeowners who invest in solar power may be in for hard times. Since 1978, when the energy crisis prompted Congress to approve federal tax credits designed to encourage investment in alternative energy, sales of windmills, solar heaters, and other devices that tap renewable power have grown to more than $3 billion a year. Critics claim that alternative energy has never been much more than a tax dodge. According to one study, in 1984 the Treasury lost $550 million in taxes due to support for renewable-energy investments.

That sum, however, is dwarfed by the $15 billion spent on the nuclear-power industry. The loss of tax benefits may well double the cost of homeowners' renewable energy systems. Although some suppliers believe that their industry is on its last legs, many contractors have diversified enough to survive.

A new burst of solar energy. Fabrizio Filosa *World Press Review* 35:53 D '88

Reprinted from *Europeo* of Milan. Solar energy machines that produce high volumes of energy are being developed in West Germany and the United States. The West German project has created a solar generator with thermochemical energy buildup that is designed to meet the energy needs of millions of Third World families. The prototype is geared to satisfying the energy requirements of an eight-member family living in a country that receives 560 watt hours per square foot of sunlight. Jurgen Kleinwachter, a physicist and co-owner of Bomin Solar, which developed the project, says that in four or five years his company will produce larger systems with the capacity to generate hundreds or thousands of low-cost kilowatt hours daily. Meanwhile, Sandia Laboratories in Albuquerque, New Mexico, has developed a photovoltaic cell that can transform an unprecedented 31 percent of solar energy into electricity.